A Highly Caffeinated Introduction to Chemical Engineering: Laboratory Manual

First Edition

William Ristenpart & Tonya Kuhl

Department of Chemical Engineering

University of California Davis

Caffæum Press

2026

A Highly Caffeinated Introduction to Chemical Engineering: Laboratory Manual

Caffaeum Press
Davis, CA
www.caffaeum.press

ISBN: 978-1-969728-01-3

Contents

Preface

This lab manual is intended to complement our textbook, *A Highly Caffeinated Introduction to Chemical Engineering.* The material herein is aimed specifically at first-year students who are majoring (or considering majoring) in chemical engineering. Accordingly, there is a much higher level of mathematical and scientific rigor compared to our general education version of this material, *The Design of Coffee: An Engineering Approach.* That general education class has no prerequisites, but for this lab manual students need some knowledge of calculus: at least one quarter or semester of differential calculus, and some integral calculus (at least taken concurrently). Some prior experience in a college-level laboratory course is helpful (but not required).

Each of the ten laboratory experiences described here is intended to serve as an experiential, hands-on learning activity to complement the respective ten chapters in the main textbook. For example, Chapter 3 in *A Highly Caffeinated Introduction to Chemical Engineering* is "Conservation of Mass," and Lab 3 here is "Process Flow Diagrams and Mass Balances for Coffee." Students are strongly encouraged to read the respective chapter *before* performing each lab. Doing so will provide a much better understanding of why certain measurements and analyses are assigned.

A main goal for the course is for students to master spreadsheet data analysis and graph preparation, skills that will help them in other classes and prepare them for more sophisticated data analyses in the future (e.g., with Python or R). Accordingly, students need access to Microsoft Excel for data analysis. Google Sheets can be used but lacks some features when plotting data with error bars. Extensive prior experience is not necessary, since gaining experience with collecting and analyzing data and communicating the results is a major component of the course.

Much of modern engineering work is performed in teams. Accordingly, the lab report questions here are designed to foster discussion and interaction with groupmates. Students should work with their groupmates to perform the measurements, analyze the data, and write lab reports summarizing their findings.

Finally, we hope that these experiments serve as an engaging introduction to the way chemical engineers think and how they analyze problems. We also hope you have as much fun performing these experiments as we did creating them!

William Ristenpart & Tonya Kuhl

Davis, California
Winter 2026

A Note about Coffee

The central theme of this lab manual involves experiments focused on coffee. Some students, however, might prefer not to consume coffee for personal reasons.

For these students, please note that you are **NOT** required to consume any coffee for any experiment outlined in this book. If this were instead a book focused on using petroleum to illustrate chemical engineering principles, there similarly would be no requirement that any student taste the crude oil. (Yuck!*) What is important is that students learn how to analyze data from laboratory experiments and communicate those results. Here we happen to use coffee, which is inexpensive and safe, yet complicated to analyze. For students who do enjoy coffee, a bonus is that they get to taste it.

If you count yourself amongst those who don't drink coffee, it would nonetheless be convenient if you team up with lab partners who do – this will make the final design contest more fun. Furthermore, if coffee is OK but you don't want caffeine, note that decaffeinated green beans are readily available, and that all the experiments described herein can be performed with decaf beans.

Lastly, if you are hesitant to taste coffee simply because you feel that you don't like its taste, please consider the possibility that you have might not have yet encountered a good cup of coffee. High quality beans, when freshly roasted and expertly brewed, can yield sublime flavor profiles with natural sweetness and notes of fruits and flowers. Plus, there are positive health aspects: brewed coffee has lots of antioxidants, which are molecules that prevent free radicals from damaging your cells. The Food and Drug Administration considers black coffee to be a "healthy" beverage when consumed at two to three 8-ounce cups per day.

So, if you're skeptical about coffee, we suggest you keep an open mind. You might be pleasantly surprised!

* Have you ever heard of "sweet" crude oil and "sour" crude oil? Sweet crude oil is low in sulfur (< 0.5%) and has a pleasant smell and slightly sweet taste. Sour crude oil contains (> 0.5%) sulfur and smells bad like rotten eggs. People actually used to taste crude oil to determine its quality… we think coffee is a much more fun liquid to taste!

Part I

Introduction

Lab 1 – Excel Tutorial, Safety Overview, and Tasting

Objectives: In this preliminary lab we will begin with a tutorial on spreadsheet data analysis. We then review important safety issues associated with hot coffee in the laboratory, and we introduce the key sensory attributes that affect the perceived quality of coffee. Finally, we collect quantitative tasting scores from everybody in the lab so that we can generate a graph summarizing how each coffee was perceived.

Equipment:

☐ Laptop or tablet ☐ Glasses ☐ Three different roasted coffees

☐ Caffeine sensory reference ☐ Citric acid sensory reference

Lab Activities:

☐ Part A – tutorial on spreadsheet data analysis

☐ Part B – safety review

☐ Part C – traditional cupping of at least two different coffees, with sensory references

Report:

☐ Signed safety sheet

☐ Column plot of entire section's tasting scores, with error bars

☐ Brief paragraph discussing key questions about the lab

Part A – Tutorial in Spreadsheet Data Analysis

The very first activity in lab will be a tutorial on spreadsheet data analysis. For this lab, *and this lab only*, you should bring your laptop or tablet where you can access Microsoft Excel (MS Office is free for UC Davis students). The teaching assistant will lead a tutorial that emphasizes the following list of topics: formulas, autofill, absolute cell references, labels & units, significant digits, scatter plots, bar plots, and calculation and insertion of error bars. You are strongly encouraged to read Chapter 1 carefully prior to the tutorial.

Spreadsheet data analysis is foundational for all the subsequent labs in this course, so make sure you understand how to perform all the calculations and spreadsheet operations. Do not hesitate to ask questions: you are here to learn, and we are here to teach you! Now is the time to master spreadsheets.

After the tutorial, put away your electronics, because it's time to brew some coffee.

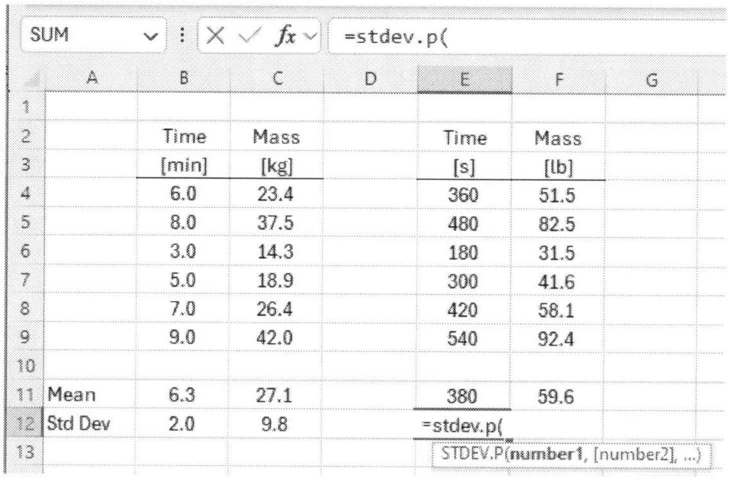

	A	B	C	D	E	F	G
1							
2		Time	Mass		Time	Mass	
3		[min]	[kg]		[s]	[lb]	
4		6.0	23.4		360	51.5	
5		8.0	37.5		480	82.5	
6		3.0	14.3		180	31.5	
7		5.0	18.9		300	41.6	
8		7.0	26.4		420	58.1	
9		9.0	42.0		540	92.4	
10							
11	Mean	6.3	27.1		380	59.6	
12	Std Dev	2.0	9.8		=stdev.p(		
13							

Part B – Safety Review

Before you perform any activities in the lab, the first crucial activity is for you to review the safety rules and expectations for the coffee lab. Your pre-lab assignment for this part is straightforward: simply read the safety rules and laboratory orientation, then sign the written safety sheet at the bottom of the next page. Submit the signed safety sheet as an image (e.g., a cell phone photo) on the course Canvas site. You don't need to submit this page; the pictorial overview is intended only to reinforce the written material on the safety sheet. If you have any questions about safety or procedure at any time, do not hesitate to ask your teaching assistant and/or instructor, or your colleagues or friends if you are performing experiments at home. Better safe than sorry!

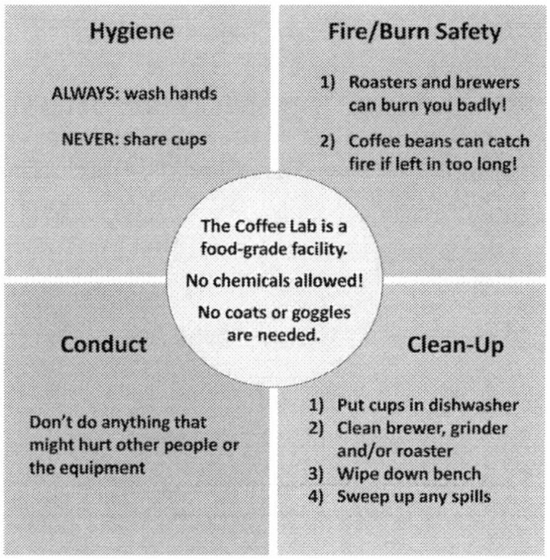

Make sure you know where a fire blanket and first aid kit are located in the remote chance that an accident occurs and you need them. Also make sure you understand what is expected in terms of cleanup after you are done.

Part C Background – Tasting Coffee

Coffee tasting is challenging. Experts spend years developing their palate to identify subtle differences in each cup. Here we provide just the basic definitions of the main sensory attributes that we will use to judge coffees during the course.

When tasting, you first smell the coffee, then loudly **slurp** some in your mouth. By slurping, you create smaller droplets (aerosols), which accomplish two goals. First, they help coat all of the taste buds in your mouth, so that your entire palate is involved in the evaluation. Second, the smaller droplets have a higher surface area-to-volume ratio, which accelerates volatilization of the aroma molecules that travel up your "retro-nasal" passageway. (A great deal of "taste" is actually "smell!") Some tasters find it helpful to "chew" the coffee in their mouth to make sure everything is maximally distributed. Finally, after you swallow, continue to assess the aftertaste. The main scoring categories are defined as follows.

Aroma – The aromatic aspects of the coffee, as detected by an initial smell before tasting in the mouth. A better fragrance receives a higher score. (In the coffee industry, some use "fragrance" to refer to the smell of the dry coffee grounds, and "aroma" to refer to the smell of brewed coffee. In this book we will use these terms interchangeably.)

Flavor – The coffee's taste character, in between the first impressions given by the first aroma and acidity to its final taste (also known as the "mid-range" notes). It is the combined impression of all the taste bud sensations and retro-nasal aromas that go from mouth to nose.

(continued on page 11)

Safety Orientation for "The Coffee Lab"

1) **Food grade facility.** The "coffee lab" is a food grade facility. Unlike many laboratories, you may eat and drink food. This means, however, that absolutely NO HAZARDOUS CHEMICALS / MATERIALS are allowed inside the coffee lab. Do not bring any chemicals inside the lab.

2) **No lab coats or goggles.** Since there are no chemicals, you do **not** need a lab coat, lab gloves, or eye protection. Regular clothing and shoes are fine.

3) **Store backpacks out of the way.** Upon entering the lab, store your backpack or bag on the designated shelving. Backpacks on the ground become tripping hazards, which is unsafe around hot water.

4) **Hygiene – washing hands.** You must wash your hands with soap and water upon entering the lab. Likewise, wash your hands after you use the restroom; after coughing, sneezing, or blowing your nose; or after doing anything else that soils your hands.

5) **Hygiene – no sharing cups.** Do not share cups with anybody else. Use sticky labels to identify your personal cup at the beginning of each lab session. Avoid the temptation to take "just a small sip" from somebody else's cup. If you think your cup is dirty or was accidentally shared, get a new one.

6) **Burn safety.** The main safety hazard with coffee is the risk of burns. The coffee pots and roasters can get hot enough to cause severe burns if contacted with bare skin. Exercise great caution around them; only handle the brewers and roasters by their designated handles. **Always pour hot coffee into a cup placed on the lab bench, never into a cup in somebody's hand.** If you do receive a burn, immediately begin running cool water over it in the sink and notify your teaching assistant and/or instructor.

7) **Fire safety.** Just like bread can catch fire in a toaster, coffee beans can catch fire in a roaster. Never leave your roaster unattended. If it begins to smoke heavily, turn it off immediately. If the beans stop moving, turn it off immediately. If the beans do catch fire, turn off the roaster and, if feasible, place a 'fire blanket' over the roaster to smother the flames. Notify the teaching assistant and/or instructor of any observed open flames, no matter how insignificant they might seem. Pay special attention to safe use of all roasters. Note the location of the room exits, the first aid kit, and the fire extinguisher.

8) **No horseplay or misconduct.** Any individual who purposely engages in unsafe behavior will be expelled from the lab and likely fail the course. Examples of misconduct include throwing things, horseplay, (wrestling, pushing, tickling, etc.), or putting anything other than coffee beans into a roaster or brewer.

9) **No unapproved equipment.** No personal brewing or roasting equipment may be brought from home into the lab. Absolutely no open flames are allowed in the lab (no camp stoves, candles, etc.)

10) **End of lab cleanup (MANDATORY).** At the end of each lab session, your group must clean your station. Place all coffee cups in the dishwasher and wash all brewing glassware you used with soap and water. Spray and wipe down your area of lab bench and use a hand-broom to sweep up any coffee grounds or beans you might have spilled on the ground. IMPORTANT: Before you leave, confirm with your teaching assistant and/or instructor that your lab cleanup is adequate to receive full points.

I affirm that I carefully read and understand the expectations for conduct and safety in the Coffee Lab.

Signature: ______________________________ Date: ____________

Print name: ______________________________________

Student ID Number: ____________________ Section number: __________

Acidity – Often described as "bright" when favorable, "sour" when unpleasant, or "dull" or "flat" when missing. (In coffee, "bright" means "good acidity!") Compare to Italian salad dressing: you want an appropriate amount of acidity from the vinegar, but too much vinegar makes it unpleasantly sour. Acidity contributes to the liveliness, sweetness, and fresh-fruit character, especially when the coffee is first slurped. A lack of acidity (a "dullness" or "flatness") receives a low score, but too much acidity ("sourness") also receives a low score. Give a high score to a "bright," lively acidity that enhances the coffee flavor.

Aftertaste – The length or duration of positive flavor (both taste and aroma) emanating from the back of the palate and remaining after the coffee is swallowed. If the aftertaste is either short or unpleasant, a lower score should be given.

Body – The tactile feeling of the coffee liquid in the mouth, or more simply, the "mouthfeel." It is highly related to the "viscosity" of the liquid. The presence of colloids, oils, and sucrose in the coffee contribute to higher body. A "watery" coffee can have good flavor but lack body, while a "thick" coffee can have strong body but bad flavor. Give high scores for pleasant body.

Balance – The overall impression of the coffee. Ideally a coffee is balanced between all the above attributes, with none dominating over any other. Think of this category as a subjective "fudge factor." If smelling and tasting the coffee overall was a positive experience, give it a high balance score. If, on the other hand, you overall had a negative experience, then give it a negative score.

Sweetness – How sweet is the coffee? When we talk about sweetness, we don't mean to add sugar to the cup! Perhaps surprisingly, certain black coffees can be perceived as having a sweetness due to the presence of naturally occurring "sweetness-associated" molecules. Black coffee is never as sweet as soda (which has a huge amount of added sugar!), but high-quality coffees can have a pronounced sweetness that is highly prized.

Part C Activities – Sensory References & Blind Tasting

Time to brew and taste some awesome coffee! For this preliminary laboratory experiment, the teaching assistant will prepare three coffees for you to taste. While you are waiting for the coffee to brew, we will begin by tasting some "sensory references." These are important in sensory and consumer science to help align tasters to use the same vocabulary for similar sensory impressions. Many people, for example, confuse bitterness and sourness when both are present in the same product, as happens with coffee. The teaching assistant will distribute two sensory references, as follows:

Bitter reference: 0.1% caffeine dissolved in water (e.g., 1 gram in 1 kg of water)
Sour reference: 1.0% citric acid dissolved in water (e.g., 10.0 grams in 1 kg of water)

To taste, pour a few milliliters of each reference into a small glass (espresso cup). Smell each one separately; do you smell anything? Why or why not? Then take a small sip of the bitter reference. Rinse your mouth with some fresh water and take a small sip of the sour reference. Repeat a few times, until you are confident you have a good sense of what is meant by "bitter" and "sour." Keep those impressions in mind as you taste the coffees. If you happen to be a "supertaster," the bitter taste will be very strong. Dilute the bitter sample by half with plain water, and by half again, to taste each dilution and see how sensitive you are to bitterness.

Tasting Data Sheet

Coffee type________________________ **Hedonic Score:** ________

Aroma: ________________________ Flavor: ________________________

Acidity: ________________________ Aftertaste: ________________________

Body: ________________________ Sweetness: ________________________

Balance & comments:____

Coffee type________________________ **Hedonic Score:** ________

Aroma: ________________________ Flavor: ________________________

Acidity: ________________________ Aftertaste: ________________________

Body: ________________________ Sweetness: ________________________

Balance & comments:____

Coffee type________________________ **Hedonic Score:** ________

Aroma: ________________________ Flavor: ________________________

Acidity: ________________________ Aftertaste: ________________________

Body: ________________________ Sweetness: ________________________

Balance & comments:____

Next, it is time to taste the coffee! For our purposes in this lab, we will perform a simple blind tasting of the three different coffees pre-prepared by the instructor. (Here "blind" means you won't know which coffee is which.) Pour a little of the sample into the cup, smell, and then taste. Feel free to discuss with your lab partners. Write down any sensory attributes that stand out to you, such as "very sour," "bitter," "fruity," "nutty," or maybe "tastes like baker's chocolate."

While the coffee is hot, your initial impressions should focus on the flavor and aftertaste. As it starts to cool, turn your attention to the acidity and the body. Next, assess the balance and sweetness. Continue to taste and adjust your evaluations as your perception of the flavors and qualities change with temperature. How do the coffees differ in their bitterness and sourness? Is one more sour or bitter? Do you like one more than the other?

Then, most importantly, you will provide a "consumer hedonic preference score" for each of the three coffees. The classic hedonic scale varies from 9 to 1 as shown. This scale was developed by military scientists in the 1950s to help improve food for soldiers, and since then it has become widely adopted by the food industry. Many chemical engineers who work in the food industry work to maximize these scores for their products!

Score	Description
9	Like extremely
8	Like very much
7	Like moderately
6	Like slightly
5	Neither like nor dislike
4	Dislike slightly
3	Dislike moderately
2	Dislike very much
1	Dislike extremely

Repeat your tasting for the other two coffee samples and record your scores and any qualitative impressions as well. It's a good idea to rinse your mouth out with water between samples. The instructor will provide a methodology for recording your scores in a group spreadsheet for the entire section. Once you're done with tasting and you have uploaded your three scores, clean up your area. Check in with your teaching assistant prior to leaving the lab.

In this lab, we only gave an overall score for each coffee. But remember the definitions of all the sensory attributes (body, acidity, balance, etc.) as you progress through the upcoming labs… your ultimate design goal is to make the best tasting coffee using the least amount of energy, and these sensory attributes are what your coffee will be judged on!

Lab Report

By your specified due date, each group will submit their lab report that includes the following.

(1) Using the tasting scores reported by everybody in your section, calculate the mean and standard deviation of the scores for each of the three samples. Then prepare a column chart showing the average tasting scores for each coffee sample. Insert vertical error bars given by the standard deviation for each respective sample. Use good graphing techniques!

(2) Write a brief paragraph summarizing your tasting notes, as well as any trends in the tasting scores. Did any samples stand out, either positively or negatively? How similar or dissimilar were your group's scores compared to everyone else?

(3) Note that the tasting scores used an arbitrary scale (1 to 9) and were based on people's opinions about how much they liked or disliked the coffee. Does this type of data count as "scientific" data? Why or why not?

Lab 1 Bonus Box – Why Do Coffees Taste Different?

Now that you've seen and smelled some green and roasted coffee beans, we should discuss a key point: coffee is a highly variable biological material! Much like the grapes used to make fine wines, coffee beans vary tremendously in taste depending on where they are grown and with how much care they are picked and processed. High quality, specialty coffee is mainly produced in tropical, equatorial zones at elevations greater than 3000 feet (equivalent to 1000 meters above sea level). Coffee "beans" are the seeds of the fruit (coffee cherry) of the *Coffea* shrub or tree. A mature plant typically produces about 20-60 pounds of coffee cherries each year, yielding between 3 and 9 pounds of green coffee beans. To maximize the quality of the green beans, the cherries must be harvested at their peak ripeness – typically bright red, yellow, or orange in color depending on the variety. Because the fruiting can happen at different times and the ripening time can vary, harvesting of optimally ripe coffee cherries is very labor-intensive with multiple picks.

Workers on a coffee farm (typically in a developing tropical country) have to care for the coffee trees, pick the ripe cherries, de-pulp the cherries to remove the fruity flesh, wash and dry the beans, sort out the bad beans by hand, bag the good beans, and then transport them to the nearest buyer – oftentimes by mule down the side of a mountain. By the time the consumer or the commercial roaster receives a shipment of green coffee beans, most of the hard work has already been done.

There are fascinating scientific aspects to the process of converting coffee cherries into usable green beans. Ultimately, the goal of processing is to separate the beans (i.e., the seeds) from the fruit and reduce their moisture content to about 12%. This enables transport and storage of the green beans for months or years with little degradation, making it easy to ship the beans all over the world. There are two main processing methods: wet or dry processing.

In dry processing the growers lay the coffee cherries out in the sun to dry, oftentimes on raised beds to facilitate airflow above and below the cherries. Over the course of 3 to 4 weeks the cherries are raked and rotated every few hours and protected from moisture to ensure even drying without mildew or spoilage. In some cases, mechanical dryers are used after a few days to accelerate the drying process. Once the cherries are dried sufficiently, the outer hull is removed to release the coffee beans. Dry processing is used in production areas with limited access to water and is more variable due to the reliance on climate conditions during the drying process. The dry process yields some of the most prized fruity flavors in coffee. It is also risky, however, since mold and spoilage can yield off flavors.

Wet processing uses a hand or mechanical depulper to remove the outer skin and some of the pulp from the cherries immediately after harvest, and then a lot of water to ultimately separate the seeds from the fruit. After depulping, there is a layer of "mucilage" (a slimy, sweet residue) on the seeds, so they are stored for a few hours to a couple of days and allowed to "ferment" to breakdown the mucilage and allow it to be removed from the seeds more easily by washing. Different fermentations occur during dry processing, but it happens in the dry fruit rather than in the liquid and mucilage exposed during wet processing.

The word "fermentation" refers to a biological process in which yeast or bacteria convert sugars into different chemicals (such as the alcohol in beer or wine). Moreover, it has long been held that a crucial step in the final flavor and aroma profile of roasted coffee is the fermentation step during the processing of coffee cherries, as the coffee beans produced via the different methods yield notably different taste profiles. Some growers and coffee processors are experimenting with various yeasts (like those used for different kinds of beer or wine) to try to enhance the flavor characteristics by better controlling the fermentation portion of the process. The impact of these experiments is still up in the air, but in the future, you might have green coffee beans fermented with an IPA yeast, or an ale yeast, or even a Chardonnay yeast. More recently, variations in the partial germination of the coffee beans (recall the "beans" are really seeds) during wet vs. dry processing steps are being studied more closely as germination may be just as, or even more, important to the final flavors. The word "germination" refers to the process by which a seed yields a plant. Given that coffee bean processing has evolved through a trial and error process - limited or impacted by the local environment (humidity and temperature), availability of water, equipment, time and resources – we hope that an even better understanding of how to produce great coffee will be available in the near future.

The key point is that both a huge amount of manual labor and a great deal of biological activity go into the preparation of green coffee beans. Despite all this hard work and the interesting scientific aspects, for the purpose of this book we focus on what happens to green coffee beans after all this processing. In other words, we treat green coffee beans as our starting material. There is nothing unusual about this; many other chemical engineering processes involve "raw" starting materials that were pre-processed in some way. Moreover, from a practical perspective, in the US it is difficult to obtain unprocessed coffee cherries.

An important point, however, must be emphasized: because coffee is a biological material, it is subject to tremendous variability. Green coffee beans vary wildly in their precise composition depending on what region they're grown in. Beans grown in Brazil are distinct from those grown in Indonesia, Ethiopia or anywhere else. Even beans from ostensibly identical coffee plants in the same plantation can also differ, depending on soil characteristics, how wet or dry the local microclimate is, or even the precise altitude of the coffee plant. (Recall many coffee plants are grown in mountainous terrain.) Moreover, the coffee cherry processing method will alter otherwise similar green coffee beans into dramatically different beans.

All these factors will influence the taste of the final brewed coffee! Perhaps surprisingly, black coffee can have amazing flavors, including berry, floral, citrus, chocolate, and vanilla. Black coffee can even have a pronounced sweet taste – without adding sugar. The image on

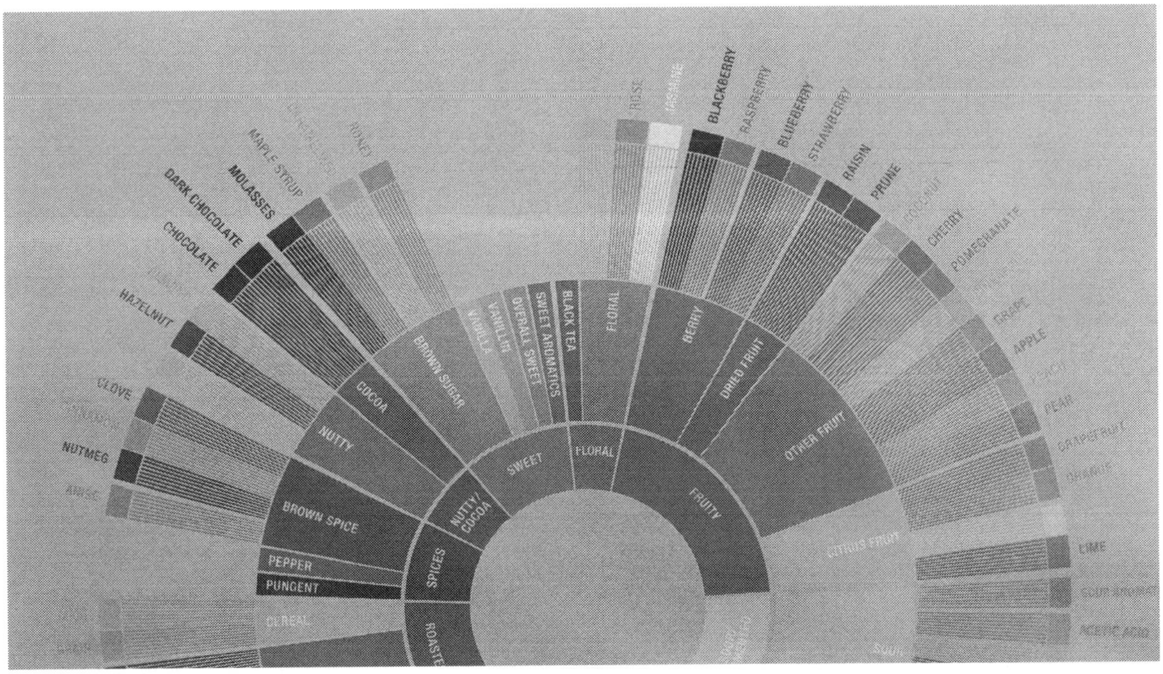

the previous page shows a part of the "Coffee Taster's Flavor Wheel," developed by a collaboration between the Specialty Coffee Association, World Coffee Research, and UC Davis. The full wheel lists 104 different flavors that have been detected in coffee!

To give an idea of the tremendous flavor profiles possible, here is an example review of a particular 2021 coffee from the Kangunu Farmers Cooperative Society in Kenya (on the east side of central Africa):

> *The fragrance from the dry grounds is boldly potent, with a strong cane sugar syrupy sweetness, and caramelized sugar notes like flambe dessert. Adding hot water, the wet aroma has mild fruited character, like tamarind chews, enveloped in caramel-vanilla sweetness, and raw sugar smells. The cup is perhaps tamer than some other Kenyas on our list, but the fruit flavors that do come into view are delicious; dried plum, Medjool date, and more of the dried tamarind. Sweetness rates very high (9.3!), and lays out complex flavors of pressed sugarcane juice, panela, unrefined cane sugar types. The cup is very 'fresh' in flavor, and while fruits are milder, they're most present in the cup, which is where we want them, right? Light roasts also have a dark black tea note that feeds in the finishing mouthfeel that is tannic with a pleasant bittering aspect. A sharp lemony accent note takes the light roasts to a level of vibrance that is uniquely Kenya!*

This review was written by Tom Owens and colleagues at Sweet Maria's, which is one of many retail distributors of specialty green beans. We urge you to do a simple web search "green coffee beans for roasting" to get a sense for the astonishing variety of green beans available for purchase. Many types cost only about $6 to $8 per pound.

Why does one batch of green coffee beans taste different from another? Everything hinges on the unique chemical composition of the green beans, and how those molecules are altered during roasting and released during brewing. There are more than 1,000 uniquely identified molecules that give rise to the flavors present in coffee, and as mentioned above, minor changes in growing conditions and processing procedure can alter the relative composition. The roasting process can enhance or diminish different aspects of the bean's flavor profile. Moreover, many of the most subtle flavors (such as "Medjool date" or "dried tamarind") are highly volatile and transient. If you wait too long after roasting, the flavors dissipate… and you're left with the more bitter molecules. Most coffee aficionados recommend that you store roasted coffee in a sealed container and brew it about 2 to 7 days after roasting. You can still brew the coffee after longer delays, but it will be more difficult to catch the types of refined and varied flavors that are possible (as highlighted in the review above).

Of course, there is more to making coffee than simply brewing it within the right time period! The upcoming labs will help you think more scientifically about the process of roasting and brewing coffee, with the ultimate goal of brewing coffee with delightful flavors like those described here.

Part II

Engineering Analysis

Lab 2 – Reverse Engineering a Drip Coffee Brewer

Objectives: In this lab we will overview and perform the process of brewing drip coffee, and learn about the "brew ratio," a dimensionless number that is crucial for coffee. We will partially disassemble a Mr. Coffee brewer to consider how it works from an engineering perspective.

Equipment:

☐ Mr. Coffee ☐ Thermometer and thermocouple ☐ Screwdriver

Activities:

☐ Part A – one brew in the Mr. Coffee, to learn how to brew and to introduce R_{brew}

☐ Part B – partial disassembly of the Mr. Coffee to reverse engineer it

☐ Part C – one more brew with temperature and volume flow rate measurements

Report:

☐ Labeled photos of the Mr. Coffee (inside and outside)

☐ Scatter plot of temperature vs. time

☐ Paragraph discussing key questions about the lab

☐ Summaries of dimensional analyses (flow rate, pressure, coffee volume)

Background

Recall there are two key components of engineering practice: *analysis* and *design*. Engineering *analysis* is the process of learning how a system of interest works, typically by combining experimental observations with theoretical interpretations. In contrast, engineering *design* refers to using that understanding to create or improve a process to accomplish some specific goal – for example, to make the best tasting coffee using the least amount of energy.

In this lab, we will be performing a specific type of engineering analysis called "reverse engineering," where we take a process that somebody else has already designed and figure out how it works. Specifically, we will reverse engineer a standard drip brewer, with a focus on the question, "What makes the water move?" Your goals in this first lab are to (1) reverse engineer a Mr. Coffee, (2) use dimensional analysis to assess some physical properties of the brewer operation, and (3) begin thinking about how material "flows" through the brewing process. This experience will prepare us for consideration of "process flow diagrams" in the next chapter.

Background – Reverse Engineering a Drip Brewer

We could just tell you exactly how a drip brewer works, but that would be boring. Instead, we want you to figure it out! To help you along the way, here are some of the items you will discover as you inspect the brewer.

- ☐ **Spring valve** – The first thing you should identify before you even open up the brewer. When you push on the spring, the valve opens; when you stop pushing on it, the valve closes. This valve allows you to pull the carafe out before brewing is complete (without making a mess).
- ☐ **Power Switch** – This is how you turn it on. It's easy to find from the outside; what does it look like from the inside? How do you think it works?
- ☐ **Wires** – As soon as you open up the bottom of the brewer you will definitely see several electrical wires inside. As you examine them, consider these two questions. (i) Why are they coated in plastic? (ii) Why are there two wires connected to the heater? Why couldn't just one wire deliver the electricity to it? (This is a subtle question!) Hint: pretend you're an electron: what path would you follow through the circuitry? On some models, you will have even more wires headed to indicator lights or other features. In a simple 4-cup Mr. Coffee brewer, one set of wires is only used to illuminate the power switch; where does the remainder go?
- ☐ **Heating tube** – There will be a large metallic "u"-shaped tube inside. Why do you think it's made out of metal? Why not plastic or ceramic? What else does this heating tube keep hot beside the water inside of it?
- ☐ **Electrical heater** – When you make electricity flow through a "resistor," the resistor heats up. This is the same effect you see in an incandescent light bulb, where it heats up so much that it starts glowing.
- ☐ **Thermal Fuse** – A thermal fuse has a special type of thin wire inside of it that allows electrical currents to pass through, but if the temperature gets too high, a component physically melts and breaks the wire, stopping the current. (Why do you think you might want a thermal fuse, **or two**, inside a coffee maker?)
- ☐ **Thermostat** – Depending on the model of coffee maker, you might have a thermostat or "Klixon" that regulates the flow of electricity. If the temperature gets too high, the thermostat or Klixon simply cuts the electricity to the heater. More expensive brewers may have a "thermistor." A "resistor" resists the flow of electricity, making the current smaller (and thus the temperature smaller). A "thermistor" is a resistor whose resistance depends on temperature. The higher the temperature, the higher the resistance, and the lower the current.
- ☐ **Check valve** – This can be difficult to locate! (Look closely inside the plastic tubes that carry the water.) A check valve, also known as a "one-way" valve, only allows fluid to flow through it in one direction. Typically, there is a flap or ball inside that opens when you push it one way but closes when you push it the other way.

Part A – Your First Drip Brew

Time to brew! First, get some roasted coffee beans (about 20 grams), and place them in the grinder. For a drip brew, you want a "medium" grind, which is not a fine powder, but more like kosher salt or coarse beach sand. (On Baratza Encore grinders this is a grind size setting of about 18.) Place a weighed mass of coffee grounds into the basket. Then pour a known mass (not volume) of cold water into the back of the drip brewer. How much water? The metric that matters is known as the "brew ratio," defined as

$$R_{brew} = \frac{\text{mass of water}}{\text{mass of dry coffee grounds}}. \tag{2.1}$$

(Pop quiz: what are the dimensions of R_{brew}?) The brew ratio is a hugely important parameter for brewing, and as we shall discover it turns out that small changes in R_{brew} can yield large changes in taste. For this initial brew, you want to use R_{brew} somewhere between 14 and 19. As an example, if you put in 20 grams of coffee grounds and 300 grams of cold water, you would have a brew ratio of 15. Remember: **the higher the brew ratio, the weaker the coffee!** (Refer to Appendix A and Appendix B for a concise summary of brewing and useful conversion factors.)

Watch the brewer carefully as it operates. Most importantly, think about the movement of the water as it moves up from the reservoir and then onto the coffee grounds. Keep this question in mind: what causes the water to move up? Pour your brew into cups and turn off and unplug the Mr. Coffee. Taste the resulting coffee. What are your qualitative impressions? Refer to the "coffee flavor wheel" – can you detect any specific flavor notes? (Refer back to Lab 1 for a guide to tasting coffee.)

Data for the First Brew

Coffee type____

Mass of water:________ *grams* Mass of beans:________ *grams*

Brew Ratio (mass of water per mass of ground beans): R_{brew} = ________

Sensory Evaluations (how does the brew smell and taste?):

____

____

Hypothesis for why the water moves:

____

____

____

Part B – Partial Disassembly of a Drip Brewer

Once the Mr. Coffee is cool enough to handle, you will partially disassemble it to figure out how it works. Refer to the instructions and the guided questions below.

1) Turn off and unplug the brewer. Remove the spent grounds and put the glass carafe somewhere to the side. After the Mr. Coffee maker is cool enough to safely handle, turn the Mr. Coffee upside-down and remove the bottom. Depending on the model, there are either screws to remove, or just plastic tabs to push in.
2) Remove the bottom plate and carefully inspect the interior of the Mr. Coffee. Try to identify each component visible underneath. Discuss with your group mates what mechanism you think there is for making the water move up to the grounds. Look at the various tubes that connect the cold-water reservoir to the upper basket where the ground coffee is placed. **IMPORTANT: Do not attempt to remove anything more or further disassemble the Mr. Coffee, this will likely break it.**
3) Measure the inner and outer diameters of the plastic tubes that you identify as transporting the water and estimate the total length of the tube from the heater to the spray head.
4) Take photos! While the Mr. Coffee is still opened up, have at least one person in your group take photos: one from the side, one from the top looking at the reservoir and grounds basket, and one looking at the inside from underneath (showing all the electrical components). These photos are necessary for your lab report.
5) After you are satisfied with your photos, review and answer the guided questions below. Work with your group to identify the key components listed on the checklist on page 19.
6) Reassemble the Mr. Coffee. Make sure that the wires are not pinched.

Part C – Second Brew with Temperature Measurements

Finally, brew a second batch of coffee with two goals: measuring the masses of water, dry grounds, brew, and wet grounds carefully, and **testing any hypotheses your group generated regarding the mechanism for how the water moved up against gravity**. You will find a thermocouple (or thermometer) helpful! The first brew was just a half pot. Automatic drip brewers don't extract well if the bed of coffee in the basket is too thin (we will study this effect in detail in later labs). This time, make a full pot with 600 grams of water. Keep the same brew ratio of 15 and see if you taste a difference! Follow these steps:

1) Carefully measure the mass of the empty carafe, ground coffee, and combined mass of the filter basket, filter and coffee, and finally the mass of water you put in.
2) As soon as the first drips start appearing, start a timer. End the timer as soon as the brew is substantially over (i.e., there is less than one drip per about five seconds). You will need this brew time for one of your dimensional analyses.
3) Measure the temperature in an appropriate location versus time during the entire brewing process. Hint: to understand why the water moves up to the spray nozzle, measure the temperature of the water dripping out of the spray nozzle. Recording every 15 seconds is recommended. What does the temperature tell you? (What is the boiling point of water?)
4) After you're done brewing, measure the mass of coffee in the carafe. Also carefully weigh the moist "spent" grounds. Did the grounds gain or lose mass during the brewing? How much drinkable coffee did you get?
5) Does the second batch taste similar to the first? If not, why not?

Reverse Engineering Guided Questions

After you remove the bottom cover of the drip brewer, carefully examine (and photograph) the interior. Recall all the important components are listed on page 19.

(i) First, trace the path of the water.

Are there any pumps or moving parts inside? ______________________

Look for a "check valve" inside of the rubber tubing. Which way does the check valve let the water flow?

____

(ii) Next, let's examine the electrical circuitry.

Where is the electrical heater? ______________________________

Why do you think it is located there, instead of where you pour the water in?

____

____

Why are there wires connected to both sides of the heater? Why can't just one wire deliver the electricity?

____

____

There is at least one "fuse" between the heater and the wires leading to the electrical plug, and there likely is a "thermostat" attached to the heater. Why do you think the fuses and thermostat are there? What purposes do they serve?

____

____

(iii) After having looked at both the path of the water and the electrical components, what do you think forces the water to move up to the spray head? Do you think it boils and floats up as steam? How would you check this? If the water is not moving up as steam, what would make the hot water move up? (Think about what happens when bubbles expand… when you vaporize liquid water, what happens to its volume?) If the check valve wasn't there, what do you think would happen when you try to brew coffee?

____

____

____

Data and Tasting Notes for Second Brew

Coffee type:____

Mass of empty carafe: ________ *grams*

Mass of water:________ *grams* Mass of beans:________ *grams* R_{brew} :________

Combined mass of plastic basket and filter paper: ________ *grams*

Mass of carafe with brewed coffee: ________ *grams*

Actual mass of brewed coffee = ________ (filled carafe) − ________ (empty carafe) = ________ *grams*

Combined mass of plastic basket, filter paper and spent moist grounds: ________ *grams*

Actual mass of spent moist grounds: ________ (filled basket) − ________ (empty basket) = ________ *grams*

Sensory Evaluations: ____

____

Temperature Data for Second Brew

Make sure the thermocouple is plugged in correctly into the meter. The big prong goes in the big hole. If the temperature goes down or is negative, it is inserted backwards! Start recording as soon as you turn on the brewer, and keep going until the brew is done.

Time (seconds)	Temperature (°C)	Time (seconds)	Temperature (°C)

(Last time recorded should correspond to 15 seconds after the brew is done.)

Lab Report

By your specified due date, each group will submit a report that includes the following analyses.

(1) Open a new file in PowerPoint and then import your best three photos of the Mr. Coffee: one from the side, one from the top looking under the lid, and one looking at the inside from underneath. Next, label all of the components and different parts of the Mr. Coffee. Use bright red arrows and easily readable text. (Hint: important labels include "heating element," "check valve" and "fuse.") Underneath each label, put a brief description (one phrase or one sentence max) of the purpose of that component or part. (Use a smaller but still legible font for the description.) You should end up with about a dozen unique labels.

(2) Use Excel to enter the measured temperature data. Generate a "scatter plot" of your temperature versus time. Make sure you label your axes, with proper units! Copy your scatter plot into PowerPoint on a separate slide. What do you think this graph demonstrates in terms of the physical mechanism for the water movement?

(3) On a separate slide, insert a text box and write a brief paragraph (10 sentences max) that clearly answers the following questions: What causes the water to move upward? Why is there a check valve? What does your temperature data indicate regarding the mechanism? Why do you think the designers of the Mr. Coffee chose this approach for heating and moving the water?

(4) Finally, perform the following dimensional analyses.

   (a) Convert your mass of brewed coffee into a volume of brewed coffee using an appropriate physical parameter. (What parameter relates mass and volume? You can approximate brewed coffee using the known value for water.) Next calculate an "average volumetric flow rate" by taking the total volume of brewed coffee and dividing by the total brew time. Using this average volumetric flow rate, use dimensional analysis to determine an expression to estimate the <u>average velocity (in centimeters per second)</u> of water moving through the tube to the sprayhead. What quantities do you need to know? Explain your logic clearly, then estimate the velocity with your data.

   (b) Using the ideal gas law and appropriate dimensional analysis, derive an expression to estimate how much volume of <u>liquid</u> water must boil to make a bubble of steam that extends all the way from the center of the heater to the top of the sprayhead. Clearly identify what must be "conserved" between the bubble of steam and the original liquid water (here "conserved" means "is the same.") Explain your logic clearly and estimate this volume of liquid water. Note that to good approximation the pressure in the steam bubble is atmospheric. Confirm that your final expression for the liquid volume is dimensionally correct.

   (c) The UC Davis coffee lab has 12 experimental stations, and typically each week there are 24 sections. If each group uses approximately the same total mass of dry coffee grounds per week that your group did for Lab 2, how many metric tons of coffee grounds are used during the academic year? (There are about 36 weeks in an academic year.) What size cubical box would we need if we stored all of the coffee grounds at once? (Specifically, what are the lengths of each side of the cube?) Finally, would this box fit in a typical dormitory room or not if the average dorm room is 16 feet long, 14 feet wide, and 10 feet tall? How much coffee is used per year in terms of "dorm room volumes?"

Lab 2 Bonus Box – Caffeine the Wonder Drug

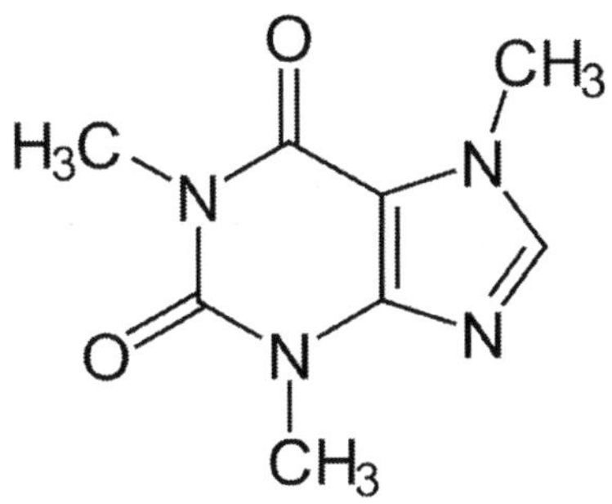

Coffee, tea, soda, chocolate - there are a lot of different ways to get your caffeine fix. But what is caffeine? In chemistry terms, it's an "organic" molecule composed of carbon, nitrogen, oxygen, and hydrogen, arranged in the molecular structure shown at right. Here we use the word "organic" in the chemistry sense that the compound contains carbon, not that it was a food grown without pesticides (as you find the word used in grocery stores). The chemical formula might look intimidating ($C_8H_{10}N_4O_2$), but caffeine is a natural substance: many different plants, including *Coffea arabica*, produce caffeine as a defense mechanism against insects – to insects, caffeine is toxic. Happily, caffeine has positive psychoactive effects on mammals: reduced drowsiness, faster and clearer flow of thought, increased focus, better body coordination.

How much caffeine do you consume when you drink a cup of coffee? Well, each 8-ounce cup of coffee has between 75 to 175 milligrams (mg) of caffeine, depending mostly on the type of coffee and how it was brewed. The roast level doesn't really affect it, since the caffeine molecule doesn't break down during roasting (it is very stable). A rule of thumb is about 100 mg caffeine per cup of coffee. Caffeine is "generally regarded as safe," since the lethal dose for humans is about 10 grams. You would have to chug about 100 cups in a row to hit that dosage!

In comparison to coffee, a typical cola (diet or regular) has a meager 25 to 45 mg of caffeine. Black tea has about 50 mg, while green tea has about 25 mg (all based on 8-ounce cups). Typical dark chocolate, say 70%, has about 80 mg per 2 ounces, or a whopping 320 mg if you ate a full 8 ounces (which is half a pound of chocolate!) Caffeine in chocolate comes from the amount of cocoa "bean" (still a seed like coffee beans) in the chocolate; the darker the chocolate, the higher the cocoa bean content and the greater the caffeine. The average adult consumes 300 mg of caffeine per day from all sources.

The range of caffeine levels in coffee stems partly from differences between the caffeine content in arabica vs. robusta coffee beans (another variety of *Coffea*). Specialty or gourmet coffees are typically arabica because it is considered to have a much better flavor profile, partially because arabica contains about half as much of caffeine and twice as much sucrose (sugar) compared to robusta. (The sucrose plays a key role in many of the roasting reactions that yield awesome flavors.) Robusta is definitely more "robust," with about 2.7% caffeine by mass to arabica's 1.5%. If you drink inexpensive instant coffee, it likely contains robusta. About 70% of coffee grown worldwide is arabica because of the better taste and higher price.

Now why is caffeine considered a "Wonder Drug?" Believe it or not, some of the most extensive research on the effects of caffeine was carried out by various branches of the armed forces. In a military situation, staying alert could be a life-or-death proposition. Compared to other stimulants, caffeine is considered one of the safest – it has extremely low incidences of abuse and adverse health effects, it is a substance most people have experience with, it is known to help alleviate sleep deprivation-induced cognitive impairment, and it even improves strength and endurance. When you drink a caffeinated beverage or take a capsule, it takes about 30 to 45 minutes to have the full effects. To overcome this delay Wrigley and the Walter Reed Army Institute of Research developed "Stay Alert" caffeinated chewing gum which reaches your system in 5 to 10 minutes – a significant advantage if you need a quick pick-me-up.

Lab 3 – Process Flow Diagram & Mass Balances for Coffee

Objectives: The overarching goal of this lab is to answer the question, "Where does the initial mass of green coffee beans ultimately end up when you make coffee?" To answer this question, we will carefully measure the mass of each material stream through both the roasting and brewing processes, to make a complete process flow diagram that includes all unit operations.

Equipment:

☐ Mr. Coffee ☐ Fresh Roast roaster ☐ Colander and brush ☐ Graduated cylinder

Activities:

☐ Part A – four brews in the Mr. Coffee, to determine mass balances and R_{abs}

☐ Part B – one roast in the Fresh Roast, to measure mass and volume changes

Report:

☐ Quantitative process flow diagram, with mass balances

☐ Summary of mass balance calculations

☐ Scatter plot of m_{brew} vs. $m_{grounds}$, with best fit slope for R_{abs}

☐ Photos of your roast showing volume change

☐ Scatter plot of SSD vs. trial slope for attempted fits of m_{brew} vs. $m_{grounds}$

☐ Paragraph discussing key questions about the lab

Background

To recap key points from this chapter: a "unit operation" is any step in a process where a chemical or physical change takes place. For example, a roaster induces a variety of chemical reactions to occur in the green beans; a grinder changes the average size of the roasted beans; a brewer extracts the hopefully tasty coffee molecules into the hot water. Each of these can be considered a "unit operation," but even within a single piece of equipment there can be multiple unit operations. A drip brewer combines the unit operations of heating, extraction, and filtration in one compact unit.

Since mass must be conserved through our process, if we put 100 grams into a unit operation, ultimately, we must get 100 grams out. This might sound obvious. Things get tricky, however, if we have multiple streams of mass moving into and out of the same unit operation. For example, a drip coffee brewer has two streams moving into it – ground coffee and cold water – but three streams moving out of it – (1) coffee to drink, (2) moist spent grounds, and (3) volatile gases (steam, carbon dioxide, and volatile organic compounds or VOCs, which are the molecules you smell in the air). The existence of these other two streams has a profound consequence: if we put in 300 grams of cold water into the brewer, we **don't** get 300 grams of coffee to drink! Some of the water is "lost" to the other waste streams.

So, if we want to design a process to make a large quantity of coffee (say 1 liter of coffee for the design competition), we must first analyze how mass flows through the system. The crucial question is: how much water and coffee grounds should we put in to get a certain amount of brewed coffee? Likewise, how much coffee should we roast?

Mass Balance for Water in Brewing

Let's focus first on the mass of water in the drip brewer. In words, our equation for conservation of mass is

$$Mass\ of\ Water\ In = Mass\ of\ Water\ Out. \tag{3.1}$$

Recognizing that we have one stream of water feed in, but three streams that contain water feed out, equation (3.1) means we have

$$m_{feed} = m_{brew} + m_{spent} + m_{evap}. \tag{3.2}$$

Here m_{feed} is the mass of cold water put into the brewer, m_{brew} is the mass of water in the brewed coffee, m_{spent} is the mass of water in the moist spent grounds, and m_{evap} is the mass of water lost as evaporated steam (water vapor) into the atmosphere. Importantly, in this balance each of these terms refers only to the mass of water in that stream, so m_{spent} is not the combined mass of solid coffee and water in the moist spent grounds – it's only the water mass.

Equation (3.2) isn't so useful yet, but it turns out we can make some very helpful **simplifying assumptions**. First, under most circumstances the mass lost to evaporation is pretty small, so we can neglect it. (You will test whether this is a good assumption in this lab.) Second, it turns out that brewed coffee is almost 99% water, i.e., the coffee solids dissolved in the brew are only about 1% by mass. So, for now, we will ignore the mass of coffee solids in the brew and just approximate it as all water mass. (As we shall see in Lab 6, another mass balance applies to the coffee solids, and we can determine the mass of coffee solids extracted into the brewed coffee.)

Third, and most importantly, we need to know how much water gets "left behind" in the spent grounds. The initially dry coffee grounds have a finite capacity to "absorb" water, in much the same way a paper towel can only absorb so much water. To a good approximation, the mass of water absorbed into the spent grounds is simply proportional to the initial mass of dry coffee grounds, i.e.,

$$m_{spent} = R_{abs} \times m_{grounds}. \tag{3.3}$$

Here R_{abs} is the "absorption ratio" that describes the ratio of how many grams of water are absorbed in the moistened grounds, per gram of initial dry coffee grounds, i.e.,

$$R_{abs} = \frac{\text{mass of water absorbed into the coffee grounds}}{\text{initial mass of dry coffee grounds}} \tag{3.4}$$

If we substitute Eq. (3.3) into Eq. (3.2), and neglect mass lost to evaporation, we obtain the very useful prediction

$$m_{brew} = m_{feed} - \left(R_{abs} \times m_{grounds}\right). \tag{3.5}$$

In other words, if we want to know **how much drinkable coffee we're going to get**, we need to know three things:

i) how much water we put in the brewer
ii) how much dry coffee grounds we put in the filter basket
iii) the numerical value of R_{abs}

A key goal of this lab is for you to measure R_{abs} experimentally, by systematically varying the mass of dry grounds but with a fixed amount of cold water. Note that equation (3.5) is a linear equation (cf. $y = mx + b$), so a plot of how much drinkable coffee you get (m_{brew}) versus how much dry grounds you used ($m_{grounds}$) should yield a straight line. Your job will be to test this hypothesis, and to determine the value of R_{abs} that best fits your experimental data.

As discussed in Chapter 3, the criterion that scientists and engineers typically use to measure the "goodness" of a fit involves finding a line that minimizes the "sum of squared differences" (SSD), also sometimes known as the sum of squared residuals. The term "residual" means the difference between an actual experimental data point and the value of the attempted best fit. If our experimental data points are denoted by y_{exp}, and if a trial best fit value is denoted by y_{fit}, then the SSD is given by

$$SSD \equiv \sum_{i=1}^{N_{tot}} \left(y_{exp,i} - y_{fit,i}\right)^2, \tag{3.6}$$

where N_{tot} is the total number of experimental data points. (In this lab you should have $N_{tot} = 4$.) The difference is squared so that negative and positive residuals don't cancel out, and we sum up all the residuals to obtain a single number that characterizes how "good" the fit is. Note that an SSD by itself is not valuable; it is only informative in comparison to other SSD's for the *same data set*. In fact, you can make the specific value of an SSD arbitrarily large or small just by changing the units on your measurement (e.g. inches vs miles)! (What are the dimensions of SSD for the experimental data in Lab 3?)

After the lab, you will perform some additional data analysis to explore the concept of SSD, to see if Excel is indeed calculating the "best fit" using equation (3.6).

Mass Balance for Roasting

To complete our process flow diagram, we also need to think about roasting. The principle of conservation of mass equally applies to roasting:

$$Mass\ of\ Coffee\ In = Mass\ of\ Coffee\ Out. \tag{3.7}$$

Just like brewing, in roasting there are multiple streams that exit the roaster. Obviously, you have the roasted beans. As you will see during your first roast, you also have the "chaff," which is the paper-like outer skin of the green coffee beans that flakes off during roasting. In small home roasters, the chaff is usually trapped in a bin of some sort by a filter screen in the exhaust from the roaster (much like the lint screen in a clothes dryer). In large commercial roasters, chaff is caught in a "cyclone separator" that uses the swirling motion of the exhaust air to direct the chaff into a bin.

Finally, the least obvious stream of mass exiting the roaster is the escaping gases. In our brewing mass balance, we neglected the volatile gases because they were so small. Roasting, in contrast, reaches much higher temperatures, and it turns out that a very significant fraction of mass is lost as water vapor, carbon dioxide, and other VOCs formed during the roasting reactions. Putting everything together, equation (3.7) becomes

$$m_{green} = m_{roasted} + m_{chaff} + m_{gas}, \quad (3.8)$$

where m_{green} is the initial mass of green beans, $m_{roasted}$ is the mass of roasted beans afterwards, m_{chaff} is the mass of chaff collected during roasting, and m_{gas} is the collective mass of all the gases that escaped during roasting.

The first three masses are easy to measure, but m_{gas} cannot be directly measured. That's ok – you will measure it indirectly in this lab!

Part A – Mass Balances for Brewing

1) First, weigh out about 65 grams of roasted coffee beans and then grind all of it together at the same time. This ensures a uniform grind size for all experiments in Part A.
2) Weigh the empty glass carafe and empty filter basket (with a dry filter paper inside of it). You will need these weights later.
3) Pour 300 grams of cold water into the Mr. Coffee and place 20 grams of your ground coffee in the filter basket. (Set the other 45 grams to the side.) Before you begin brewing, however, weigh the entire Mr. Coffee and record the mass. **Make sure the electrical cord is unplugged and draped over the top of the Mr. Coffee**; otherwise, your mass measurements will vary with how much tension is on the cord. You will eventually compare this initial mass to the mass after the brewing is done. What will these numbers tell you about the mass of lost gases (like steam)?
4) After the Mr. Coffee is done brewing, weigh the entire mass of the Mr. Coffee in the same fashion you did at the start (in step 3), and then separately weigh the glass carafe (with coffee) as well as the combined mass of the plastic basket (with moist spent grounds). This data will let you do a careful conservation of mass calculation.
5) Go ahead and taste the coffee – how does it taste?
6) Repeat steps (3) and (5) for three more brews (no need to repeat step 4, unless you want to). Each time, use exactly 300 grams of cold water, but vary the mass of ground coffee beans. Recommended: second brew, use 30 grams; third brew, use 10 grams, fourth and final brew, use 5 grams. Make sure you carefully record the mass of brewed coffee each time, you will need this data to determine R_{abs}. How does the brew ratio affect the taste?

Data for First Brew (and Flow Diagram)

Coffee type: ____

Mass of empty glass carafe: ________ *grams*

Combined mass of plastic basket and filter paper: ________ *grams*

Mass of cold water:______ *grams* Mass of beans:______ *grams* R_{brew}=_______

Mass of entire Mr. Coffee (with water and grounds) **before** brewing: __________ *grams*

Mass of entire Mr. Coffee (with water and grounds) **after** brewing: __________ *grams*

Mass lost to evaporation: ________ – ________ = _______ *grams*

Data for First Brew (continued)

Mass of carafe with brewed coffee: _______ *grams*

Actual mass of brewed coffee: ________ – ________ = _______ *grams*

Combined mass of plastic basket, moist filter and spent moist grounds: _______ *grams*

Actual mass of spent moist grounds: ________ – ________ = _______ *grams*

Sensory Evaluations: ____

____

Data for Second Brew

Mass of cold water: _______ *grams* Mass of beans: _______ *grams* R_{brew}=_______

Mass of carafe with brewed coffee: ________ *grams*

Actual mass of brewed coffee: ________ – ________ = _______ *grams*

Sensory Evaluations: ____

____

Data for Third Brew

Mass of cold water: _______ *grams* Mass of beans: _______ *grams* R_{brew}=_______

Mass of carafe with brewed coffee: ________ *grams*

Actual mass of brewed coffee: ________ – ________ = _______ *grams*

Sensory Evaluations: ____

____

Data for Fourth Brew

Mass of cold water: _______ *grams* Mass of beans: _______ *grams* R_{brew}=_______

Mass of carafe with brewed coffee: ________ *grams*

Actual mass of brewed coffee: ________ – ________ = _______ *grams*

Sensory Evaluations: ____

____

IMPORTANT ROASTING ADVICE

If you're using a Fresh Roast roaster, here are some important guidelines.

Fan	Always start the fan at 9. As the beans lose mass, you may choose to decrease the fan speed after a few minutes, but start at 9 to make sure the beans circulate. Never go below a fan speed 6! Always turn the fan setting back to 9 during the cooling cycle.
Power	A good power setting to start with is 9, but you can decrease this to extend roasting times. Below 6 does not work well. Minimizing energy here is tricky. Lower power equals longer roasting time, and that (may) require more total energy than a shorter roasting time. (We study this in detail in Lab 5.)
Time	You can start with 9-minute roast times, but the initial time setting is less important because you should watch your roast carefully. You can always stop the roast (by pressing the 'cool' button) or add more time if needed. The roasting time is typically 6 to 8 minutes, not including the automatic 3-minute cooling cycle.

Safety note: Don't put too many green beans into the Fresh Roast roaster – and don't put too few! If you put in too many beans they will not circulate properly and they will catch fire. Likewise, if you put in too few, they will get too hot too quickly and likely catch fire. **The max is about 120 grams** of green coffee beans. Do not leave your roaster unattended for any reason. If you see lots of smoke, stop the roast! To stop it, press "Run/Cool." Make sure fan speed is 9.

Part B – Mass Balances for Roasting

1) After you feel adequately caffeinated, it's time to turn to roasting. Review the safety guidelines for proper use of the roaster. Carefully weigh out 120 grams of green coffee beans, and place them in the graduated cylinder – what volume do they occupy? Take a picture of your green beans in the graduated cylinder, using a white piece of paper for background.

2) Do a medium roast, using the recommended settings for your roaster (see the advice above). The last 3 minutes of the roast is a cooling cycle, so if you set a time of 6 minutes, the total time in the roaster will be 9 minutes. **Observe the beans as they roast!** Proper roast time is highly dependent on the beans' moisture and the specific roaster, which can vary considerably from one roaster to another. Listen for the first crack and **observe the color** – this will tell you more than the timer! How to tell if you've reached a medium roast? Refer to the images on the back cover, a medium roast is close to a "deep brown."

3) Next, measure the mass of an empty metal bowl. This is where you are going to collect the chaff, which is the papery material that flakes off the beans as they roast. Do not tare the bowl. The mass of chaff is small and if you tare the bowl, you may not register a mass change. After the roast cooling cycle is complete, carefully remove the **hot** top of the Fresh Roast and place it in the metal bowl. Most of the chaff is inside. **After the top has cooled** carefully open the top and brush the chaff into the bowl.

Notice: this lab is the only time you will need to collect and weigh the chaff. **In all future labs, you will simply discard the chaff into the compost,** with no need to separate any loose chaff from the roasted beans.

4) Next, dump the roasted beans into the metal mesh colander while holding it over the metal bowl to catch loose chaff. Important: for this lab you need to weigh all chaff! (In subsequent labs we won't collect the chaff; in fact, chaff is full of antioxidants that are good for human nutrition, so there is no need to remove it.) Carefully collect all the chaff, including any remaining loose chaff in the roasting chamber.

5) Weigh the collected chaff (subtract the empty bowl mass). Weigh the roasted coffee beans. Is the chaff weight appreciable or negligible? (Note many scales have a minimum weight tolerance – the chaff by itself won't register unless you use a heavy enough container.)

6) Pour the roasted beans back into the graduated cylinder. What volume do they now occupy? How much have they expanded? Take a picture of the roasted beans in the graduated cylinder. What type of roast would you characterize your beans as? Smell the roasted beans. The fragrance is likely not that strong yet, but it will develop over the next couple days. Feel free to taste one (they're edible.)

7) Once the roasted beans are completely cool (less than body temperature), place them into a storage bag. Warm beans can partially melt the bag causing unpleasant flavors – be patient! Before putting them in the bag, use a permanent marker pen to write your group's name, section number, type of coffee, and the mass of green and roasted coffee. In future labs you will also record energy usage.

Don't put warm beans into the plastic bag! The plastic will melt and ruin your beans (unless you like molten plastic in your coffee!!!) Let them cool until they are below your body temperature, i.e., they feel cooler than your skin.

Roasting Data

Coffee type____

Mass of green beans:________ *grams* Volume of green beans:_________ mL

Roaster settings: ________________ Roasting time: ______ *mins* Color: _____________

Mass of roasted beans:________ *grams* Volume of roasted beans:________ mL

Mass of empty metal bowl:________ *grams* Mass of bowl with chaff:________ *grams*

Mass of chaff: ________ – ________ = ________ *grams*

Mass lost as volatile gases: ________ – ________ – ________ = ________ *grams*

Note and observations: ____

____

____

Lab Report

By your specified due date, each group will submit their lab report that includes the following material.

(1) Create a simple "coffee process flow diagram" for roasting and brewing. Represent each unit operation in the process as a simple box with inlet and outlet streams (cf. the laundry example in Chapter 3). Using the data from your first brew, insert your group's quantitative mass measurements (numerical values). We will focus here only on the coffee solids and water streams – be sure to include waste streams. Make sure you include appropriate units (e.g., grams) with each number you insert! Start with green coffee beans and end with a final cup of coffee. Pretend that the coffee you roasted in Lab 3 is also what you brewed in Lab 3 (it's not, but our goal here is to make a representative process flow diagram). You will find that you roasted more beans than you could use in one brew, which is OK. Just indicate that the difference goes to "storage," so that your masses balance out.

(2) The second page should show your calculations to obtain the mass measurements used in your process flow diagram. What you measured in lab should be clearly identified, and your calculations should be easy to follow. It is OK if you hand write this page and then scan or photograph it to insert into PowerPoint, but you will probably find it easier to use Excel (cf. step 4).

(3) On the next page of your report, insert your "before/after" photos of the green and roasted beans. Clearly label them and show your calculations of percent volume change.

(4) In Excel, use your experimental data to generate a scatter plot of m_{brew} versus $m_{grounds}$. Use the LINEST function (NOT the "insert trendline" option) to make a "best-fit" line and determine the slope and intercept. How does the slope of this line relate to R_{abs}? What should the intercept be?

(5) We will now explore the concept of "least squares analysis" by calculating some SSDs. Arrange your experimental values of m_{brew} in a column; these correspond to y_{exp} in equation (3.6). In the next column, use the values of the slope and intercept calculated in the previous formula to calculate y_{fit} for each distinct value of $m_{grounds}$. Then use equation (3.6) to calculate the SSD. Repeat this process for at least 10 other trial values of the slope (holding the intercept constant), ranging from a trial slope that is 50% smaller than the best fit slope to one that is 100% larger (twice as large). Clever use of functions and autofill will make this task very straightforward. Finally, make a scatter plot of SSD versus trial slope. What shape is the scatter plot? Was the best fit slope truly the minimum value? Make sure you label your axes with *correct units* (what are the units on SSD here?)

(6) On a separate final slide, clearly answer the following questions:

(i) What numerical value did you get for R_{abs}? Is it better to use the slope, or a single brew experiment to determine R_{abs}? Why? If you start with 600 grams of water, how much drinkable coffee would you get if you used 50 grams of ground coffee?

(ii) How much water was "lost" to steam or vapor during the brewing? Is the water lost to the gas phase a significant or negligible fraction of the overall water mass balance? In other words, how good was our first simplifying assumption on page 37?

(iii) How much mass of initial green coffee beans do you "lose" during roasting? What fraction is the chaff? Where does the rest of the mass go? Do you think this answer will be the same for light and dark roasts?

Lab 3 Bonus Box – What happens to spent coffee grounds?

Every time you brew coffee, you are left with two things: hopefully the first is a great cup of coffee; the second is spent coffee grounds. We all know what to do with a great cup of coffee, but what can be done with spent coffee grounds? You definitely don't want to rebrew them, because you have already removed about 2/3 of the soluble coffee from the grounds (that's the 18 to 22% percent extraction that you will learn about in Lab 6). What's left behind are less desirable flavor molecules that take longer to extract and will not enhance your next brew.

If you go online, there are a wealth of resources and ideas for uses of spent coffee grounds. First, spent coffee grounds are a great fertilizer, especially if your soil is alkaline (pH>7). Most plants prefer soil pH to be between 6 to 7. Acid-loving plants include cucumbers, eggplant, carrots, roses, and hydrangeas (which happen to be pH sensitive and turn a lovely purplish blue when fertilized with coffee). The grounds are also thought to keep some insects (ants, slugs, and snails) away and even the neighbor's cat from using your garden as a litter box. Spent coffee grounds also can be used in your kitchen refrigerator or freezer as a deodorizer and are a handy way to keep worms alive for fishing or aid their digestion in the compost heap. But the uses don't stop there: spent grounds can even be part of your beauty regimen as an exfoliate (body scrub), face mask, or added to some shampoo to help remove styling product residue from hair.

These sorts of uses are great at the home scale where you might generate 100 grams of spent coffee grounds each day. Your local coffee house generates 10 to 50 pounds per day, though, and coffee product manufacturers can generate thousands of pounds of spent coffee grounds every day. For example, one medium size Bay Area cold brew coffee company we like produces 15,000 to 20,000 pounds of spent coffee grounds – per day! To put this on an even grander scale, Americans consume about 400 million cups of coffee per day, which corresponds to a staggering 24 million pounds of spent coffee grounds, each and every single day of the year.

Unfortunately, most of the spent grounds end up in landfills. Just as you might find it difficult to get every possible use out of your spent coffee grounds, coffee houses and companies have a difficult time too. That Bay Area manufacturer offers their spent coffee grounds for free to anyone who can find a way to use them. Unfortunately, few accept their offer. This is a shame, because the spent coffee grounds contain useful chemicals not completely removed during brewing – such as remaining aromatic flavor compounds, lipids and fats, proteins, minerals, and phenolic antioxidants, which could be removed and used in other products. The bulk of the material, however, is cellulosic (plant cell walls). Cellulose and hemicellulose can be used as feedstocks for biorefineries, but with crude oil (petroleum) relatively inexpensive, biorefining of coffee grounds isn't profitable. Simply burning the spent grounds for energy is difficult as the grounds must be dry and burning them generates a lot of particulates, which are not good for air quality.

The engineering challenge is to develop an environmentally sound and economically viable process to better utilize spent coffee grounds – transforming a waste stream to a more valuable product. As more research is done, new treatments, processes, and uses may become available. Indeed, a completely integrated use/exploitation processing system where every valuable component is extracted and used optimally is possible with existing technology – and with further engineering work, may someday be economically viable.

Lab 4 – Transient Mass Balances and pH in Drip Coffee

Objectives: In this lab, we will consider the drip coffee brewer as a dynamic system, and we will perform experiments to determine whether the pH (the acidity) varies with time because of evaporative mass loss or chemical reactions.

Equipment:

☐ Clever Coffee Dripper ☐ Electric kettle ☐ Mr. Coffee ☐ pH meter ☐ Roaster

Activities:

☐ Part A – Mr. Coffee brew with systematic sampling of the mass and pH over 60 minutes

☐ Part B – a Clever Coffee brew, to compare the pH of your light roast from Lab 3

☐ Part C – two roasts in the Fresh Roast (one dark, one light) for the next lab

Report:

☐ Scatter plot of mass in the carafe vs. time

☐ Scatter plot of pH vs. time and corresponding $[H_3O^+]$ vs. time

☐ Estimates of the *rate* of acidification via evaporative loss vs. chemical reaction

☐ Tasting notes and pH of light roast vs. dark roast

☐ Paragraph discussing key questions about the lab

Background

Imagine that you carefully roast some high-quality green coffee beans, and a few days later you carefully brew them. You taste it, and you think it's absolutely the best coffee you've ever made. Excited, you text your friend and invite them over. Half an hour later your friend arrives, but the response when they taste your coffee is a polite "I guess it's OK." Confused, you taste it again, and you agree… now it tastes mediocre. What happened?

A key thing to realize about coffee is that the taste of any given brew will change with time. The moment the hot water hits the coffee grounds, three processes start occurring. The first process is "mass transfer," which is the migration of the molecules out of the solid phase (ground beans) and into the liquid phase (water). This process of extraction is manifested as an obvious change in the color of the water, from transparent to dark. (We examine mass transfer in more detail in Lab 6.) This process stops once you separate the grounds from the brew.

Two other processes, however, continue to occur. The second process is the escape of volatile organic compounds (VOCs) from the solid and liquid phases into the gas phase. These VOCs comprise the wonderful aromas that you smell while brewing coffee; over 1000 unique different molecules have been identified as contributing to the aroma of coffee.

As long as roasted or brewed coffee is exposed to the open atmosphere, VOCs will continue to volatilize and escape. The VOCs in the roasted or brewed coffee are eventually depleted, so the wonderful coffee aroma disappears, and the flavor stales and suffers.

The third process involves acidification. A common observation is that the longer a carafe of coffee sits on a hot-plate, the sourer it tastes. The sensory perception of 'sour' reflects the presence of acidic molecules in solution (recall our citric acid sensory reference in Lab 1.) Some acidity is good: coffee aficionados highly value coffees that yield a pleasant acidity, or "brightness," in their coffee. (Here the term "brightness" has nothing to do with how it looks visually!) Too much acidity, however, makes the coffee taste increasingly sour. These are the primary reasons that coffee sitting on a hot plate for a long time tastes "stale" and "sour."

The increase in acidity in coffee is readily measured with a pH meter. Recall that pH is a measure of the acid concentration in a solution, defined as

$$\mathrm{pH} = -\log_{10}[\mathrm{H_3O^+}], \tag{4.1}$$

where $[H_3O^+]$ is the concentration, in moles/liter, of "hydronium ions." A hydronium ion is basically a water molecule with an extra hydrogen atom attached to it, giving it a net positive charge. Note that the pH is a logarithmic scale. Distilled water has a pH of 7, which means the hydronium concentration is 10^{-7} mol/L. In contrast, the vinegar you buy at the store has a pH of about 3, so the hydronium ion concentration is 10^{-3} mol/L. The difference between pH 3 and pH 7 might seem small numerically, but it's a logarithmic scale: the vinegar is 10^4 = 10,000 times more acidic.

What determines the pH of coffee? The initial pH upon brewing will depend on several factors, including the variety of beans, how they were roasted (light vs. dark), as well as the initial pH and "alkalinity" of the water you use to brew. If the initial pH of the water is slightly basic (say around 8), then the initial pH of the brewed coffee might be around 6. With time as the brewed coffee sits on a hot-plate, however, we typically observe a significant decrease in pH (i.e., increase in acidity), associated with an increasingly sour taste. If the coffee reaches a pH of 5 from the original pH of 6, your coffee is now 10 times more acidic than when first brewed – with a very noticeable effect on taste!

This observation raises an important question: ***why* does the acidity increase with time?** There are at least two hypotheses to consider.

(1) **The slow distillation hypothesis.** First, note that the carafe is an "open system," in the sense that the liquid is exposed to the open atmosphere. This means that the water and other volatile components can evaporate and escape – leaving the acidic species behind in the liquid. In this sense, the carafe on the hot plate is like a slow, simple distillation (without the condenser). Since the amount of water in the system decreases, and the acidic species remain constant, the acid concentration increases and corresponding pH decreases.

(2) **The slow chemical reaction hypothesis.** The second point to note is that brewed coffee has hundreds of different chemical species. Although in Chapters 3 and 4 we focused on non-reacting systems, there is no guarantee that a real system is non-reacting. A possible hypothesis is that some chemical species in solution undergo chemical reactions that produce hydronium ions – thus increasing the acidity (decreasing the pH).

The main goal of Lab 4 is for you to perform careful measurements of the brew mass and pH vs. time to assess which hypothesis best agrees with the experimental observations.

Towards that end, some brief words about chemical reaction rates are necessary. We will consider chemical reactions in much more detail later in this book, but for now let's note that a natural question is, "How fast does the chemical reaction happen?" When we ask that question, what we really mean is "How many acid molecules per liter of brew are produced per minute?" If we ignore the mass lost to evaporation, the coffee pot is what chemical engineers refer to as a "batch reactor." If you neglect the mass loss, and assume all the observed change in hydronium concentration is due solely to chemical reactions, then you can use the pH data to get an estimate for the rate of reaction as

$$\text{rate} = \frac{d[H_3O^+]}{dt} \approx \frac{\Delta\,[H_3O^+]}{\Delta\text{ time}} \approx \frac{\text{change in }[H_3O^+]}{\text{change in time}} \approx \frac{[H_3O^+]_{final} - [H_3O^+]_{init}}{t_{elapsed}}, \tag{4.2}$$

where "init" means the initial hydronium concentration (at $t = 0$), "final" means your final concentration, and $t_{elapsed}$ is the amount of time (in minutes) until your final measurement. Here the square brackets mean "concentration in moles per liter." Note that the rate has units of moles per liter per minute, i.e., mol/(L×min). Also note that the above expression is only a rough estimate (because we approximate the time derivative), but that's OK if we only need an estimate... we call it an "engineering approximation."

As an example, let's say you measure a coffee with initial pH = 6.1, and an hour later the pH is 5.2. The corresponding hydronium concentrations and approximate reaction rate are

Initial: $$[H_3O^+] = 10^{\wedge(-6.1)}\,\frac{\text{mol}}{\text{L}} = 0.00000079\,\frac{\text{mol}}{\text{L}} = 7.9\times10^{-7}\,\frac{\text{mol}}{\text{L}}, \tag{4.3a}$$

Final: $$[H_3O^+] = 10^{\wedge(-5.2)}\,\frac{\text{mol}}{\text{L}} = 0.0000063\,\frac{\text{mol}}{\text{L}} = 6.3\times10^{-6}\,\frac{\text{mol}}{\text{L}}, \tag{4.3b}$$

Rate: $$\frac{d}{dt}[H_3O^+] \approx \frac{6.3\times10^{-6}\frac{\text{mol}}{\text{L}} - 7.9\times10^{-7}\frac{\text{mol}}{\text{L}}}{60\text{ min}} = 9.2\times10^{-8}\,\frac{\text{mol}}{\text{L min}}. \tag{4.3c}$$

Don't be fooled by the seemingly small numbers: the human tongue can definitely detect the acid associated with such small concentrations! But the big question for this lab is whether we need to invoke chemical reactions at all, or if the observed change in pH is consistent with acidification due to evaporation of the water.

Part A – Measuring the pH versus time with dark roasted coffee

Important: you need to multitask during this experiment – get part A started, and then simultaneously start working on part B then part C.

1) Get ready to make some coffee in the Mr. Coffee, using some **very dark** roasted coffee. The darker the better. If you use light or medium roasted coffee, the initial brew is so acidic that it is difficult to measure a change in pH. Make sure you use dark beans.

2) Prepare a large amount: use about 600 grams of water, and 40 grams of coffee ($R_{brew} = 15$). Measure the pH of the water that you put into the brewer.

3) As soon as it is done brewing (i.e., there are no more drips) make note of the time by starting a cell phone timer: the end of the drip brewing process is our $t = 0$. Pour out **small** samples to taste (about half an ounce max) and assess the perceived acidity (or "brightness") of the coffee. How does it taste?

Special warnings about handheld pH meters:

1) The only button you should push is the "POWER" button.
2) DO NOT push the "CAL" button (this changes the calibration!)
3) DO NOT submerge the buttons in coffee or water – only the tip!
4) Always measure room temperature coffee, less than body temp.
5) Rinse and gently blot dry the pH probe tip between readings.

4) After your initial taste tests, carefully measure the mass of the brew in the carafe. This is your starting mass ($t = 0$), and you will measure both the mass in the carafe and the pH every 5 minutes for the next hour.

5) Next, review the warnings below about proper usage of handheld pH meters. They typically only provide an accurate pH when the coffee is cool – don't try to measure the pH of hot coffee! To do this, pour a little bit of coffee in a glass mug (about a quarter ounce), let it cool down for a few seconds, then pour it into a second glass mug to further cool it. Ideally the brew is quickly cooled to body temperature or less (i.e., not hot to the touch). Then pour the cooled coffee into the pH probe cap to measure, insert the pH probe, and measure the pH. Wait until the reading stabilizes (about 10 seconds), then record the pH of the brew sample.

6) **Very important:** pour the pH measured brew back into the carafe!

7) For the next 60 minutes, **keep the glass carafe on the hot plate, with the Mr. Coffee turned on,** and take a carafe mass measurement and small sample for pH measurement every 5 minutes. (Keep your cell phone timer running.)

8) After the final pH and mass measurements, feel free to taste as much coffee as you like. How has the perceived acidity changed with time compared to the initial brew?

Part B – Measuring the pH of your medium roasted beans

1) Our next task will be to brew your medium roast beans from Lab 3. We will use a new brewing technique – the Clever Coffee Dripper. Grind 20 grams of your medium roast and place the grounds into a conical filter paper placed in the Clever Coffee Dripper.

2) Do not put the Clever Coffee on a cup yet! It has a spring valve that only activates when placed on a cup or mug. Leave the Clever Coffee on the countertop for brewing.

3) Use the electric kettle to heat water to 94°C and pour in about 360 grams of hot water using a swirling motion to fully wet the grounds. Start a timer when done pouring. Brew for about five minutes.

4) After five minutes of brewing, place the Clever Coffee onto a mug or beaker to filter and drain the brewer.

5) Measure the pH of the brew. How does it compare to the dark roast?

6) Finally, do a "blind" taste of your medium roast vs. the dark roast that you let sit on the hot plate for an hour. Let somebody else pour them for you so that you don't know which is which. How do they taste? Which do you prefer?

Leave the Clever Coffee on the countertop, not a mug, while brewing. Use a "swirling" motion while pouring to fully wet all the grounds. If desired, put the lid on to minimize heat loss while brewing.

Mass and pH versus Time Data

Coffee type____

Mass of water:________ *grams* Mass of beans:________ *grams* R_{brew}:________

Mass of empty carafe:________ *grams* pH of tap water: ___________

Time (minutes)	Mass (grams)	pH
___________	___________	___________
___________	___________	___________
___________	___________	___________
___________	___________	___________
___________	___________	___________
___________	___________	___________
___________	___________	___________
___________	___________	___________
___________	___________	___________
___________	___________	___________
___________	___________	___________
___________	___________	___________
___________	___________	___________

Sensory Evaluations at $t = 0$ **(end of brewing):**

____

____

Sensory Evaluations at $t = 60$ **minutes:**

____

____

pH Data for the Clever Coffee Medium Roast

MEDIUM ROAST Type: ______________________________

Mass of hot water: _______ *grams* Mass of grounds: _______ *grams* R_{brew}:________

Grind: ________ Extraction time: ______ *minutes* Water temperature: _______ °C

pH of brew: __________ Sensory Evaluations: ______________________________

____

____

Part C – Roasting a lighter and a darker roast

During the 60 minutes of your pH data acquisition, we will multi-task to also perform two roasts – a lighter roast and a darker roast, both with the same type of green beans. Use the exact same procedure as last week, but don't bother measuring the chaff or the volume change (those activities were only for Lab 3). Observe the beans as they roast! For the lighter roast, use a shorter roasting time and/or a lower power setting than for the darker roast. Note: you can always hit the Run/Cool button to stop heating and go to the cooling cycle early while roasting. Make sure you let the roaster cool for about five minutes between roasts. Label and store your beans for next week. Remember, only bag fully cooled beans.

If you like, you can taste the roasted beans (just pop one in your mouth and chew). Can you taste a difference between the lighter and darker roasts?

Roasting Data

Coffee type: ____

First Roast (Lighter)

Mass of green beans: ________ *grams* Mass of roasted beans: ________ *grams*

Roaster settings: ________________ Roasting time: ______ *mins* Color: ____________

Notes: ____

Second Roast (Darker)

Mass of green beans: ________ *grams* Mass of roasted beans: ________ *grams*

Roaster settings: ________________ Roasting time: ______ *mins* Color: ____________

Notes and observations: ______________________________________

____

____

Lab Report

By your specified due date, each group will submit their lab report that includes (1) a plot of brew mass versus time, (2) plots of pH and $[H_3O^+]$ vs. time, (3) calculations assessing the two hypotheses for the acidification, and (4) tasting notes and pH of light vs. dark roast, and (5) a brief paragraph discussing your group's observations and assessments.

(1) First, use Excel to generate a scatter plot of the brew mass versus time. Use linear regression to insert a best fit line, and calculate the average value of the evaporative mass flowrate M_{out}. What overall percentage of the initial mass was lost?

(2) Next, use Excel to generate scatter plots of the pH versus time and $[H_3O^+]$ vs. time. Use linear regression to insert best fit lines in each. How good are the fits?

(3) Perform calculations to assess the validity of each hypothesis for why the brew acidifies.

a. First, assume that there are no chemical reactions. Using the multicomponent mass balance methodology discussed earlier in this chapter, calculate what the pH *should be* at the final time given the initial pH and the measured mass loss over the course of the experiment. (Hint: what is the molarity of water?) Assume that the mole fraction of acidic species in the evaporative stream is zero (i.e., assume the acids are completely non-volatile). How different or similar is this calculated pH to the experimentally observed pH?

b. Next, assume that the mass loss is negligible and treat the carafe as a batch reactor. What was the average reaction rate over the entire experiment? Examine your scatter plot of $[H_3O^+]$ vs. time. If the changes in $[H_3O^+]$ were due entirely to chemical reactions, did the *reaction rate* increase with time, decrease with time, or stay constant with time?

c. Which of the two hypotheses do you think is most likely correct? Using that mechanism, if you let the coffee sit on the hot plate for three times as long (say 3 hours instead of 1 hour), and if we assume the evaporative rate or reaction rate stayed constant (likely poor assumptions!), what would the final pH be?

(4) Next, write down your tasting notes for your lighter roast Clever Coffee brew versus your darker roast drip brew (both initial and final) and their pH values. Qualitative taste impressions are fine! Which did each groupmate prefer?

(5) On a final separate page, write a brief paragraph that clearly answers the following questions:

(i) Why do you think the pH decreases with time? Use quantitative reasoning from your analyses in part (3) above.

(ii) What sources of experimental error do you think might have affected your measurements? Could they have altered your conclusions?

(iii) Can you propose a modification to the experiment procedure here to more accurately assess which proposed mechanism for acidification is operative?

(iv) How much did the taste of the coffee vary with time as it sat on the heating plate? Qualitative impressions are fine! Did you notice a correlation with pH?

Lab 4 Bonus Box – What Happens During a Roast?

Roasting is a critical step towards obtaining "nirvana in a cup" quality coffee. There is a myriad of physical and chemical changes that occur when you roast green coffee beans, all triggered by heating the beans. Initially, green coffee beans contain about 12% water moisture, less if they have been stored for a long time. As the beans and water inside them heat up, the water vaporizes. At a certain point, a bit above 200°C, the vapor pressure in the bean, due to the expansion of gases with temperature, is so large that the beans actually crack open. An audible crack or popping noise can be heard in most roasters. The cracks and fissures in the beans allow the water vapor and other gases to escape and also cause a significant increase in the volume of the bean. In the image above it is possible to see some of these cracks.

The same type of cracking happens when you pop popcorn. The corn kernels contain water and gases that heat up, and once the pressure exceeds the strength of the corn kernel shell – they explode. The cell walls of coffee beans are just tougher and less continuous, so that explosive event is less dramatic. Moreover, have you ever noticed that old corn kernels don't pop up that much? The reason is simply that the kernels have dried out and there isn't enough water in the kernel to build sufficient pressure to pop the kernel. If you add some water to the old bag and let it sit overnight, those kernels will absorb the moisture and pop much better the next day. As you keep heating coffee beans they even crack a second time starting around 230°C. This further opens up the bean and further breaks down the plant cell walls. As a result, oils in the bean migrate to the surface and cause the bean to appear shiny or oily.

But pops and cracks aren't all that is happening. Just like baking or roasting other foods, there are a host of complex chemical reactions that also occur as the beans heat up. The most obvious changes are from "Maillard" reactions, which result in the brown color of the roasted bean. Maillard reactions occur between proteins and sugars – just like when you toast bread. The other main type of chemical reaction is "pyrolysis," which is a general term that means a reaction caused by heat in the absence of oxygen – which occurs inside the bean. Pyrolysis caramelizes sugars and carbohydrates and changes the fats in the bean to aromatic oils. If you have cooked food, you have carried out both Maillard and pyrolysis reactions, but hopefully not too much combustion! Yes, if you go too far with your coffee roast, you can set the coffee beans on fire, just like wood in a fireplace. Combustion reactions are caused by heat, but in the presence of oxygen. Note that combustion is not good in coffee!

Another important aspect of green coffee beans is that they contain a lot of chlorogenic acids, which make up 4-9% of the green beans. Why focus on this one type of molecule? Well, chlorogenic acids are not found in a lot of plants, but they are incredible antioxidants. Drinking coffee is actually a great source of antioxidants! Roasting, however, breaks down chlorogenic acids; in fact, only about 40% of these molecules remain in a medium roast and really dark roasts are below 10%. So, lighter roasts are "brighter" (more acidic due to chlorogenic and other acids) and may actually be healthier.

If all these changes that occur during roasting are confusing for you to follow, you are not alone. There are so many different possible reactions during roasting that scientists still don't have a full handle on all that is going on either. Moreover, the composition of the green beans varies like any natural crop, so a roast that gives you "nirvana in a cup" using one type of green coffee beans may lead to brown swill with another. But all is not lost: try different roasting levels and explore what yields the best tasting brew.

Lab 5 – Measuring the Energy Used to Make Coffee

Objectives: The overarching goal of this lab is to answer the question, "How much energy does it take to make coffee?" We will measure the heat capacity of water, and compare the energy requirements for brewing, roasting, and grinding.

Equipment:

☐ Electric kettle ☐ Kill-a-Watt meter ☐ Fresh Roast ☐ Clever Coffee Dripper

Activities:

☐ Part A – heating water in the electric kettle (twice), to determine C_p for water

☐ Part B – measuring the energy of grinding and brewing using Clever Coffee Drippers

  ☐ Measure the average wattage used during grinding, brew light vs. dark roast

  ☐ Blind taste test of dark versus light roast from Lab 4

☐ Part C – measuring the energy of roasting

  ☐ Wattage measurements of the cooling fan (no beans)

  ☐ Wattage and temperature measurements vs. fan and power settings (no beans)

  ☐ Energy and temperature measurements during roasting (with beans)

Report:

☐ A scatter plot of energy vs. temperature for heating water, and table of C_p calculations

☐ Three scatter plots for roasting wattage and temperature versus power and fan settings

☐ Scatter plots of energy vs. time and temperature vs. time for roasting

☐ Pie chart of energy per gram used by different unit operations

☐ Paragraphs discussing tasting notes and key questions about the lab

Background – Energy and Power Measurements

As discussed in Chapter 5, the First Law of Thermodynamics can be expressed in words as

> "*Although energy assumes many forms, the total quantity of energy is constant, and when energy disappears in one form it appears simultaneously in other forms.*"

In a course on thermodynamics, you study the implications of the First Law in much more detail. Here we focus on understanding how to characterize the amount of thermal energy, i.e., "heat," that is transferred during the process of brewing and roasting coffee.

As you have now seen, in the process of making coffee there are three major unit operations that require electrical energy: (i) heating of water for brewing, (ii) roasting, and (iii) grinding the roasted coffee. When you plug in an electric roaster or an electric hot water kettle,

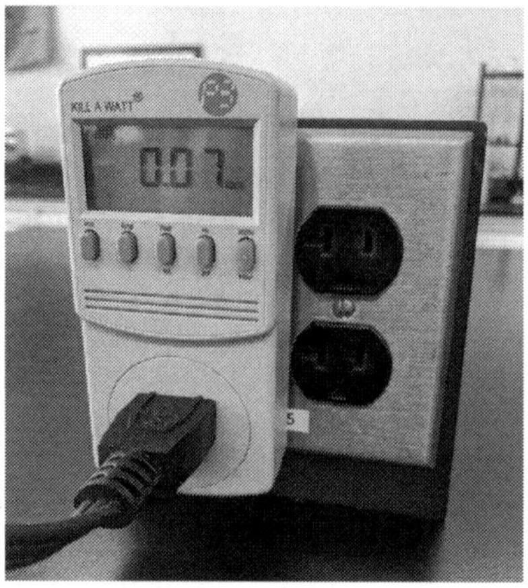

the energy of the moving electrons is transferred to heat through a "resistive element." Essentially, the moving electrons generate some heat each time they collide with a molecule inside the resistor that doesn't allow them to move so easily. The higher the electric current (i.e., how many electrons per second), the more collisions occur, and more energy is transferred to heat. The "resistive element" in the automatic drip coffee maker in Lab 2 was a copper wire embedded underneath the metal u-shaped tube. In an electric roaster, there is typically a heating coil similar to ones used in hair dryers.

Again, the higher the electrical current (i.e., how many electrons per second), the more collisions occur, and more energy is transferred to heat. Referring to the First Law of Thermodynamics, mathematically we can write

$$\Delta U - Q = 0, \tag{5.1}$$

because in the heater we don't do any mechanical work. The change in internal energy is given by $m\, C_p\, \Delta T$. The heat generated in the resistor is equal to the electrical energy that passes through it, i.e.,

$$Q = \Delta E_{electrical}, \tag{5.2}$$

where $\Delta E_{electrical}$ is the change in energy of the electrical current going through the resistive heater (and what is measured by the electrical meter).

If we know how much energy we apply to the electrical resistor, how much hotter will the surrounding water get? Balancing the two terms yields

$$\Delta U = m\, C_p\, \Delta T = \Delta E_{electrical}. \tag{5.3}$$

Hopefully, this expression is intuitive. If you try to heat a larger mass, more energy is needed; if you try to raise the temperature by a lot, then more energy is needed. Precisely how much energy depends on the specific heat capacity of that material, which varies with the chemical composition of the material. As discussed near Table 5.1 in Chapter 5, many metals (like copper or gold) have pretty low specific heat capacities, around 0.3 J/(g °C), but water actually has a very large specific heat capacity, around 4.2 J/(g °C). However, you don't have to take our word for it: you'll be measuring C_p for water yourself in this lab. How? The key idea is to measure the electrical energy transferred to the kettle as a function of temperature for a known mass of water; a proper linear regression will then yield a line with a slope related to the heat capacity of water.

What about the energy needed for roasting coffee beans? This energy is much more complicated, because it involves vaporization of water from the green beans, and also many different chemical reactions – some of which are endothermic (meaning they require energy) and some of which are exothermic (meaning they release energy). Despite this complexity, we can still determine how much energy a roaster uses with some simple measurements in an empty roaster (no coffee beans at all). A Fresh Roast roaster has two primary energy-using components: the fan and the heating element. We will measure their energy usage independently versus the respective fan and power settings, to see how the arbitrary scales for the roaster settings actually translate into watts.

A very important and counterintuitive concept is that *lower* fan speeds typically yield *hotter* inlet air temperatures inside the roaster. A major goal of this lab is for you to corroborate this result, and to figure out why it is true – and what the implications are for obtaining a good dark roast!

Importantly, the electrical meters we use measure not only the cumulative energy used (in kW-hr), but also the instantaneous power usage (in Watts). For several experiments you will need to measure the power versus time, and then "numerically integrate" with respect to time (as described in section 5.8) to get the cumulative energy used. Mathematically,

$$E = \int P(t)dt \approx \sum P(t_i) \cdot \Delta t_i, \tag{5.4}$$

where P is the power, in watts, measured at some specific time. For example, if you measure the power every 10 seconds for 300 seconds, then an approximate value of the total energy used is simply the sum of the instantaneous power measurements each multiplied by ten seconds. This procedure is sometimes called "quadrature," and the approximation for the integral becomes increasingly accurate as Δt_i shrinks towards zero.

Part A – The Energy to Heat Water

1) First, add a known mass of water to the kettle (about half full); make sure you record the mass of water. Plug in the kettle through a zeroed Kill-a-Watt meter, but don't turn the kettle on yet. You zero a Kill-a-Watt meter simply by unplugging the meter from the outlet and plugging it back in. **Press the purple button to get kW-hr.**

> **Make sure you press the purple button on the meter to measure *energy* in kW-hr!** Later in this lab you will push the "watts" button to instead measure watts.

2) Set the kettle for 94°C. Don't start it yet! Get a cell phone timer ready.
3) Once you're all set, turn on the kettle. Starting from $t = 0$ and $E = 0$, record the time and temperature whenever the energy reading on the meter increases by 0.01 kW-hr.
4) Discard the hot water, then repeat the measurements one more time. Use a different mass of water, with the kettle almost full (record this new mass). **Make sure you zero your Kill-a-Watt meter** and rinse the kettle a few times to cool it down.

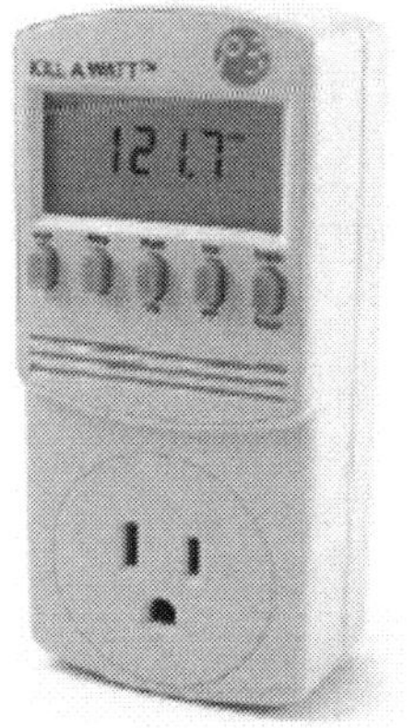

5) Again, once you are set, record the time, temperature and energy at every 0.01 kW-hr increment. Later you will graph and analyze these data, but discuss with your lab mates – qualitatively how did the energy usage vary with temperature and time for the two different masses?
6) Don't discard the full kettle of hot water – you're going to use that water in part B.

Temperature and Energy vs. Time Data: Half-Full Kettle

Mass of water in **HALF-FULL** kettle: _________ *grams*

Time (sec)	Temp. (°C)	Energy (kW-hr)	Time (sec)	Temp. (°C)	Energy (kW-hr)

Temperature and Energy vs. Time Data: Almost Full Kettle

Mass of water in **ALMOST FULL** kettle: _________ *grams*

Time (sec)	Temp. (°C)	Energy (kW-hr)	Time (sec)	Temp. (°C)	Energy (kW-hr)

Part B – The Energy of Grinding (and Taste Comparison of Lighter vs. Darker Roast)

1) Set the electric kettle to the side momentarily. It is now time to use the Clever Coffee Dripper for two head-to-head brews with your roasts from Lab 4. But first: we are going to measure the energy of grinding.

2) Weigh out a mass of your lighter roast appropriate for a Clever Coffee brew (e.g., 20 grams). Set the grinder to an intermediate grind size (around 18 on a Baratza). Make sure the grinder is plugged in to the Kill-a-Watt meter. **IMPORTANT:** set the Kill-a-Watt meter to measure watts, not kW-hr.

3) Start a timer and begin grinding your lighter roast. The power in watts will fluctuate a bit as you grind, so do your best to estimate the average wattage. Record both this average wattage and the time it took to grind.

4) Repeat this process for your darker roast, using the same mass of beans and the same grind size setting. Record the average wattage and the time needed to grind. Did the darker roast require more energy, less energy, or the same amount of energy to grind compared to the lighter roast?

5) Set up two Clever Coffee Drippers to brew. You will be doing a taste comparison, so try your best to prepare the brews identically: make sure the brew ratio is the same, the water temperature the same, and the extraction time the same.

6) Now, it's time for a blind taste test! Have one person in your group be the server and have the other(s) turn around and/or close their eyes. Serve out samples to your groupmates into two different small cups, but don't let them know which is which. Taste each one. Can you determine which is your lighter roast and which is your darker roast? Refer to the coffee flavor wheel – what flavor notes do you detect? Which one tastes "best?" Write down your sensory impressions of each brew.

Energy Data for Grinding

Mass of **LIGHTER** roasted beans: _________ *grams*

Average power during grinding: ________ *watts = joules/second* (It's ok, it fluctuates!)

Time spent grinding: ________ *seconds*

Energy usage: ________ *watts* × ________ *seconds* = ________ *joules*

Energy usage in kW-hr: ________ *joules* $\times \frac{1\ kW-hr}{3.6 \times 10^6\ joules}$ = ____________ *kW-hr*

Mass of **DARKER** roasted beans: ________ *grams*

Average power during grinding: ________ *watts = joules/second* (It's ok, it fluctuates!)

Time spent grinding: ________ *seconds*

Energy usage: ________ *watts* × ________ *seconds* = ________ *joules*

Energy usage in kW-hr: ________ *joules* $\times \frac{1\ kW-hr}{3.6 \times 10^6\ Joules}$ = ____________ *kW-hr*

Data for Brewing Blind Taste Test

Grind size: __________ Water temperature: ________ °*C*

Mass of hot water: _______ *grams* Mass of beans: _______ *grams* R_{brew}: ________

Extraction time: ________ *minutes* *(These should all be the same for both brews!)*

Coffee type and roast level: ____

Sensory Evaluations / Blind tasting notes: __________________________________

____

Coffee type and roast level: ____

Sensory Evaluations / Blind tasting notes: __________________________________

____

Part C – The Energy of Roasting

The next activity is to perform a detailed analysis of energy usage during roasting. There are three goals: (i) to measure the wattage used by the fan during cooling (with no heating), (ii) to measure the wattage used by the roaster vs. the power setting, and (iii) to measure the wattage and temperature of the air versus fan speed at constant power setting. **IMPORTANT:** you will do all three of these experiments with an **EMPTY** roaster, no beans. After you complete these measurements, you will do two actual roasts with beans.

1) Plug the roaster into the Kill-a-Watt meter and set it to measure watts (not kW-hr). Do NOT put any beans in the roaster: the following experiments use an **empty roaster**.
2) What is the power draw, in watts, when the roaster is plugged in but not running? (This is known as the standby power or "vampire power.")
3) Set the fan speed to maximum (9 on a Fresh Roast), start the roast, and immediately start the cooling cycle. This turns the heater off and allows you to **measure the energy used just by the fan**.
4) Wait about 10 seconds for the wattage to stabilize, then record the wattage.
5) Decrease the fan speed by one increment (e.g., to 8 on a Fresh Roast). Again, wait about 10 seconds for the wattage to stabilize, and then record the wattage.
6) Repeat this process for fan settings 7, 6, and 5. (No need to investigate fan settings 1 through 4 on a Fresh Roast because they typically aren't strong enough to circulate beans.)
7) Next you will measure the **energy usage when the heating element is on**. Set the fan speed to maximum and the power setting to half minimum (5 on a Fresh Roast). Start a timer and begin the "roast" (again, there should be no coffee beans in the roaster at this point). At thirty seconds, record both the wattage and the temperature in the roaster (on a Fresh Roast, rotate the knob quickly to display temperature in °F).
8) Increase the power setting by one increment (e.g., to 6 on a Fresh Roast) and restart your timer. At thirty seconds, record the wattage and temperature.

9) Repeat this process until you have wattage and temperature data at every power setting up to maximum (9 on a Fresh Roast).

10) Next, you will keep the power constant at maximum (9) and change the fan setting. Start with the fan setting at maximum (9). After waiting 30 seconds, measure and record both the wattage and temperature.

11) Decrease the fan speed by one increment (e.g., to 8) while keeping the power setting constant. After 30 seconds, record the wattage and temperature.

12) Repeat this process until you reach the half maximum fan speed (5 on a Fresh Roast). Each time, record the wattage and wait 30 seconds to record the inlet air temperature.

13) Finally, it's time to roast some coffee! Weigh out some green beans (120 grams each) and perform two roasts. Use different green beans than in prior labs, so that next time you can taste a new type of bean. During the roast, keep track of time and carefully note when you hear "first crack." For both roasts, record the instantaneous watts and inlet temperature periodically versus time (about every 30 seconds). You will need this for your quadrature calculations!

Roaster Power Data

Average power on standby: ________ *watts*

Wattage for the Fan Only (*Cooling cycle, no beans*)

Fan Setting	Power (watts)	Fan Setting	Power (watts)	Fan Setting	Power (watts)
________	________	________	________	________	________
________	________	________	________	________	________

Wattage vs. Power Setting (*Max fan speed, no beans, after 30 seconds*)

Power Setting	Temp. (°F)	Power (watts)	Power Setting	Temp. (°F)	Power (watts)
________	________	________	________	________	________
________	________	________	________	________	________
________	________	________	________	________	________

Wattage vs. Fan Setting (*Max power, no beans, after 30 seconds*)

Fan Setting	Temp. (°F)	Power (watts)	Fan Setting	Temp. (°F)	Power (watts)
________	________	________	________	________	________
________	________	________	________	________	________
________	________	________	________	________	________

Energy Data for Roasting Coffee Beans

1st roast Coffee type: ____

Mass of green beans: ________ *grams* Mass of roasted beans: ________ *grams*

Roaster settings:_____________ First crack: ______ *mins* Color: ________________

Time spent roasting: ________ *minutes* Total energy usage: ________ *kW-hr*

2nd roast Coffee type: ____

Mass of green beans: ________ *grams* Mass of roasted beans: ________ *grams*

Roaster settings:_____________ First crack: ______ *mins* Color: ________________

Time spent roasting: ________ *minutes* Total energy usage: ________ *kW-hr*

1st roast			**2nd roast**		
Time (sec)	**Temp. (°F)**	**Power (watts)**	**Time (sec)**	**Temp. (°F)**	**Energy (watts)**
________	________	________	________	________	________
________	________	________	________	________	________
________	________	________	________	________	________
________	________	________	________	________	________
________	________	________	________	________	________
________	________	________	________	________	________
________	________	________	________	________	________
________	________	________	________	________	________
________	________	________	________	________	________
________	________	________	________	________	________
________	________	________	________	________	________
________	________	________	________	________	________
________	________	________	________	________	________
________	________	________	________	________	________
________	________	________	________	________	________
________	________	________	________	________	________

Lab Report

By your due date, each group will submit their lab report that includes the following: (1) a scatter plot of energy versus temperature; (2) a spreadsheet table of your C_p calculations for water; (3) scatter plots of your roaster wattages and temperatures; (4) a pie chart showing the energy per mass for each unit operation; (5) numerical power integrations; and (6) a brief discussion.

(1) Enter your electric water kettle data (the time/energy/temp measurements) into Excel and prepare a scatter plot of cumulative energy (in kW-hr as the vertical axis) versus temperature (as horizontal axis). You should have two curves on this plot, for the half-full and full kettles. Use different marker symbols and colors for each curve. Make sure the scatter plot has a legend that specifies which set of points corresponds to which mass of water.

(2) Next, prepare a table in Excel that has 4 columns: mass, overall ΔT, overall total energy, and C_p. Here "overall" refers to the change between the beginning and end of the trial. Enter the two sets of data (one for each mass), and then calculate C_p in the fourth column using an appropriate formula. (Which is what? Refer to Chapter 5.4.) Report your measured heat capacity in terms of $J/(g \times °C)$.

(3) Now, for the roaster data. Prepare the following set of five scatter plots.

  a. The power in watts versus the fan setting during the cooling cycle.
  b. The power in watts versus the power setting, at max fan speed.
  c. The power in watts versus the fan setting, at max power setting.
  d. The inlet air temperature versus the fan setting, at max power setting.
  e. The instantaneous power in watts versus time when roasting coffee beans.

(4) Perform numerical integration of the power versus time for the two roasts to obtain the cumulative kW-hr. How much do these values differ with the kW-hr from the meter?

(5) Prepare a pie chart that compares the energy usage for the three different unit operations you measured: the kettle, the grinder, and the roaster. For each operation, calculate the energy required per unit mass of material: per gram of water for the kettle, per gram of green coffee for the roaster, and per gram of roasted coffee for the grinder. Make sure your pie chart is clearly labeled, and insert labels that specify the numerical value of **kW-hr per gram** for each operation.

(6) On a final separate page, write clear answers to the following questions.

  (i) How does the slope of the energy vs. temperature depend on the mass of water? How did your measured C_p for water compare to the established value? Be quantitative – what is the percent difference? What might account for any discrepancy?

  (ii) What qualitative impressions (tasting notes) did you have about the lighter roast compared to the darker roast?

  (iii) On a per gram basis, how much energy does grinding require compared to roasting? Express your answer in terms of a percentage. And what used more energy per gram: heating water or roasting coffee? How much more? Be quantitative.

  (iv) What trends did you observe in the wattage usage within the roaster? How much more power does the heater use versus the fan? Why does the inlet air temperature in the roaster *decrease* as the fan speed increases? (Refer to your scatter plot 3d, above.) What settings should you use to achieve darker roasts? Do you think it's a good idea to use a very low fan speed? Why or why not?

Lab 5 Bonus Box – Water: The Fluid of Life (and Coffee)

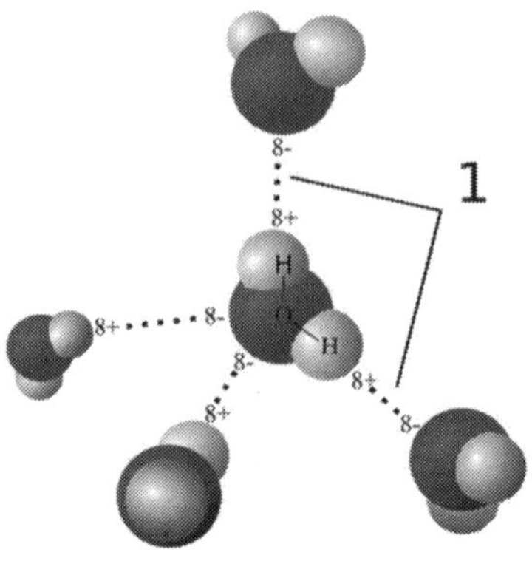

Water: the fluid of life? Yes, water is pretty unique, and frankly special. Water is one of the most important fluids on the planet – and the universe. First, and most obviously, we couldn't survive without water. In the context of coffee, water is almost 99% of what we are drinking in the cup. But that is just the starting point for why water is so interesting.

In terms of its chemical structure, water is H_2O – two hydrogen atoms bonded to an oxygen atom. The figure at right shows the sort of structure or orientation of water molecules relative to each other in ice, where the dark sphere represents oxygen and the lighter spheres are the hydrogens. Because oxygen has what's known as "lone pair electrons" and is very "electronegative" (which simply means that it attracts electrons), each water molecule can form "hydrogen bonds" with surrounding water molecules (as depicted by the dashed lines). All the details of this are pretty cool, but the main aspect of this (if you aren't interested in the details) is that water forms hydrogen bonds with itself and consequently has a certain structure. Why do we care? Well, because of this structure, when you freeze water the molecules are able to make those hydrogen bonds really well – so that each water molecule is hydrogen bonded to 4 other water molecules – and they, therefore, make a nice crystal structure. This ordered crystal formation is why we have the beauty of snowflakes like the one shown here.

In liquid water, the molecules can move around more so there are actually less than 4 hydrogen bonds per water molecule. As a result, the water molecules can actually be a little closer. These differences in packing are why ice floats on liquid water – ice is less dense than liquid water because the hydrogen bonds keep the molecules a little further apart on average. The lower density of ice also means that the volume of water actually expands upon freezing. Most liquids contract and decrease in volume when they freeze or solidify. Pretty cool, but a note of caution – this is why you shouldn't freeze water in a sealed bottle or container. If you put a can of soda in the freezer, the can will bulge and potentially burst, leaving a slushy mess.

In terms of life, ice being less dense than liquid water is pretty important if you think about bodies of water like oceans or lakes. If ice were denser than liquid water, bodies of water (like lakes) would freeze from the bottom and then continue freezing all the way up, killing the plant life and beaching the animal life on a giant, solid ice cube; less plants, less fish, less animals, less life.

Another interesting point regarding the need for each water molecule in ice to make 4 hydrogen bonds is that this can't happen at the very surface of ice. At the surface there isn't water surrounding all of the molecules; there is air on one side. Air can't hydrogen bond. Because of this, the surface molecules of water are not as strongly bound and are fluid, which enables you to ski, snow board, and ice skate. The slipperiness of that thin liquid water layer at the surface enables your skies, board, or skates to easily slide due to the fact that you are actually skidding along liquid water. In fact, the surface layer of ice doesn't actually freeze until about -10°C or 14°F. At this point you are literally on solid ice and the friction goes up dramatically. Of course, not too many people are interested in trying to ski, snow board, or skate when it's that cold (brrrrr!)

These changes from solid to liquid are captured in the "phase diagram" of water, which tells you at what temperature and pressure water is in its various phases like solid (ice), liquid (water), and gas (vapor). The large arrow shows what happens at 1 atmosphere (that's normal atmospheric pressure at sea level). The horizontal dashed line is 1 atmosphere of pressure and the vertical lines show the melting or freezing point and the boiling point of water. As you move horizontally to the right along that dashed line, below 0°C, water is in the form of ice, as you heat and move to the right the water melts at 0°C turning into liquid water, and if you keep heating it will start to boil at 100°C.

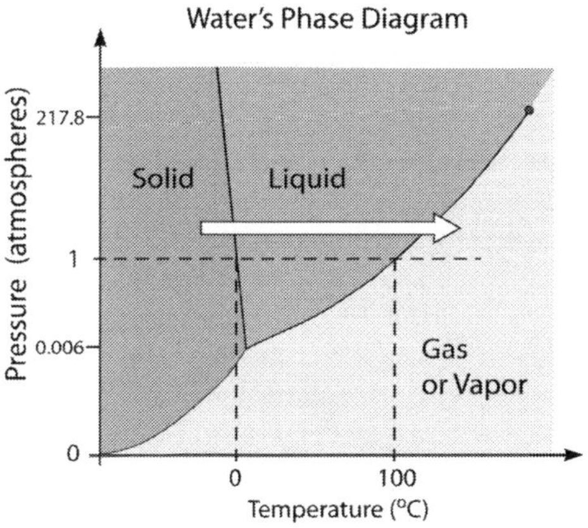

If you like to go hiking up in the mountains or live near Denver, you might have noticed some differences when you cook. A big reason for these changes can be understood from the phase diagram of water. As you go up in elevation, atmospheric pressure decreases – that's what makes it harder to breathe. At an elevation of 8000 feet (1.5 miles above sea level) the pressure is only about 0.75 atmospheres, or 25% lower than the pressure at sea level. Now, if you look back at our phase diagram, the lower pressure means that water actually freezes at a slightly higher temperature and water boils at a lower temperature - about 92°C instead of 100°C. That's why you have to cook pasta longer when you are backpacking (the boiling water is not as hot up there), and the instructions for baking a cake are different at high elevations vs. at sea level. At the top of Mount Everest (with elevation of 29029 feet or 5.5 miles) the pressure is only 0.33 atmospheres and water boils at a little above 70°C.

How does this matter for coffee? Well, another cool thing is that below a certain pressure – around 0.006 atmospheres, which you can reach pretty easily with a vacuum pump – when you heat ice it doesn't melt to a liquid. Instead, it goes directly from solid ice to vapor. This process of transitioning from a solid phase directly to a gas phase is called "sublimation," and is easy to observe with dry ice (made of solid frozen carbon dioxide). If a material (like coffee) has water in it and is frozen, the sublimation process will remove the water without melting it. You can freeze the coffee then put it under vacuum, sucking all the moisture out as water vapor at a much lower temperature. This "freeze drying" process is used not only to make expensive instant coffee but also to process and store various pharmaceuticals that would degrade if they were left in liquid water or heated. Almost half of the instant coffee sold in the United States is freeze dried. Spray drying is used for the rest where small coffee droplets are sprayed into heated air to evaporate the water. Freeze drying is a little more expensive than spray drying but is thought to better preserve the aroma and flavor molecules. To make both methods more economical, the brewed coffee is first concentrated using a variety of techniques before drying. (This point is discussed more in the Lab 12 Bonus Box.)

Many coffee aficionados look down on instant coffee, and it is true that instant coffee is not as popular in the United States as elsewhere. Around the world, however, instant coffee is actually about 40% of the global coffee market – which is a lot of coffee, made possible by the phase diagram of water!

Lab 6 – Mass Transfer and "Flux" in Brewing

Objectives: In this lab we will study the concepts of mass transfer and flux as applied to the extraction of solid coffee grounds into water. We will compare quantitative measurements of "total dissolved solids" with qualitative sensory evaluations of the taste.

Equipment:

☐ French press ☐ Electric kettle ☐ Digital refractometer

☐ Kill-a-Watt meter ☐ Roaster ☐ Paper cups ☐ Pestle and mortar

Activities:

☐ Part A – measuring TDS and PE with several Clever Coffee brews

  ☐ Three brews at different grind sizes

  ☐ Two brews at different temperatures

  ☐ One long brew, with periodic samples to get TDS vs. extraction time

☐ Part B – one or two roasts to have beans for next time

Report:

☐ Labeled column plot of TDS vs. brew condition

☐ Labeled column plot of PE vs. brew condition

☐ Scatter plot of TDS vs. extraction time

☐ Paragraphs discussing tasting notes and key questions about the lab

Background – Brew Strength & Extraction

What do you do if you brew coffee that is too "weak" or too "strong?" We intuitively know that if the brew is too strong we should increase the brew ratio or grind the coffee beans a little coarser. If you have a brewing method where you control the time for brewing, you might decrease the extraction time. This laboratory will provide a fundamental understanding of mass transfer and why these sorts of adjustments dramatically alter the quality and "strength" of the brew. Because there are so many different types of molecules in brewed coffee, it is challenging to measure the concentration of any specific molecule. Instead, we measure the cumulative concentration of *all* the different molecules that move from the solid coffee grounds to the liquid. This cumulative concentration is known as the Total Dissolved Solids (TDS) and is often expressed as a mass percentage. In brewed coffee, a typical TDS is about 1%, which means that 1% by mass is dissolved solids, and close to 99% is water (with the remainder as trace emulsified oils and gases).

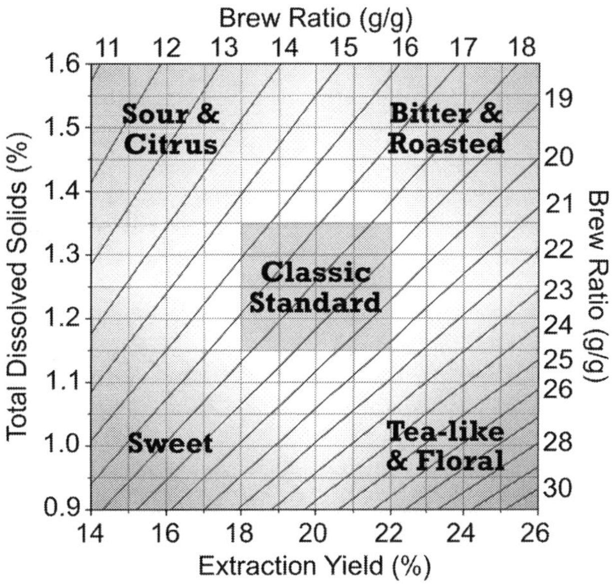

Usually when people refer qualitatively to how "strong" a particular cup of coffee is, they are responding to their perception of the TDS. It is easy to visualize that 'weak,' mostly translucent coffee has a relatively small TDS, while 'strong,' extremely dark coffee has a large TDS. This perspective, however, is incomplete. A second independent variable governing the quality of the brew is the "percent extraction" (PE), which refers to the percent weight of solids originally in the coffee grounds that were transferred to the liquid phase. (The PE is also known as the "yield" or "extraction yield.") In other words, the PE is how much of the solid coffee mass was removed from the grounds to the water. Typical coffee grounds are composed of about 70% cellulose-like materials and other compounds that do not dissolve in water. They are insoluble. The maximum possible PE is thus approximately 30%.

It is undesirable, however, to extract all 30% of the soluble compounds. Taste tests have consistently found that coffee extracted to the full 30% is unpleasantly bitter. In contrast, 'under-extracted' coffee at less than about 15% PE is unpleasantly sour and vegetal. The broad consensus is that coffee should be extracted to within the range 18% < PE < 22%, with a TDS somewhere near 1.2 to 1.3% (this is the region labeled "Classic Standard" above). These results are summarized qualitatively in the "coffee brewing control chart" shown above, which relates the TDS (on the vertical axis), the PE (on the horizontal axis), and the brew ratio (on the diagonal lines) with sensory aspects of the brewed coffee.

Note that the sensory attributes in the corners of the control chart indicate where careful sensory descriptive analysis indicates that those attributes are *maximized*. For example, coffee has perceptible bitterness across the entire range of values in the chart, but if you want to maximize bitterness, you want a high TDS and a high PE (the upper right corner of the chart). In contrast, if you want to maximize perceptible sweetness, then you should aim for lower TDS and lower PE values (the lower left corner). The coffee will still have perceptible bitterness and sourness in that corner, but there should be more perceptible sweetness compared to the other regions of the chart. (See the "Further Reading" in the appendices for more details regarding how this chart shown was developed.)

Given that the flavor depends sensitively on the PE, a natural question is: how do you measure the PE? There is no easy way of directly measuring PE, but you can calculate it indirectly using the same sort of mass balance that we discussed in Lab 3. The main idea is the same: the mass of coffee molecules fed into a unit operation must be equal to the mass of coffee molecules that exits the unit operation, i.e.,

$$Mass\ of\ Coffee\ Solids\ In = Mass\ of\ Coffee\ Solids\ Out. \quad (6.1)$$

Recognizing that we have one stream of coffee solids moving in, but two streams that contain solids moving out, equation (6.1) means we have

$$m_{dry\ grounds} = m_{spent\ grounds} + m_{coffee\ solids\ in\ brew}. \quad (6.2)$$

We emphasize that each of the terms in equation (6.2) represents only the mass of coffee solids in that stream, so $m_{coffee\ solids\ in\ brew}$ is just the mass of coffee solids in the brew, while $m_{spent\ grounds}$ represents *just* the mass of coffee solids remaining in the spent grounds (*not* the combined mass of water and solids in the spent grounds). Accordingly,

$$TDS \text{ in } \% = \frac{m_{coffee\ solids\ in\ brew}}{m_{brew}} \times 100, \tag{6.3}$$

and the percent extraction is defined as "how much of the original solids were removed into the liquid,"

$$PE \text{ in } \% = \frac{m_{coffee\ solids\ in\ brew}}{m_{dry\ grounds}} \times 100. \tag{6.4}$$

Combining, we obtain

$$PE = TDS \times \frac{m_{brew}}{m_{dry\ grounds}}. \tag{6.5}$$

Note that because TDS and PE are usually expressed as a percentage, we multiply by 100 in equations (6.3) and (6.4) to convert them from mass fractions. Equation (6.5) works either way, provided you consistently use either mass fractions or percentage values for **both** TDS and PE (if TDS is in %, you obtain PE in %). Note that we can also derive equation (6.5) from equation (6.4), but the process is a little less obvious. In this case, using mass fractions we note the mass of coffee solids in the brew is obtained by measuring the TDS where $m_{coffee\ solids\ in\ brew} = TDS \times m_{brew}$ and $m_{spent\ grounds} = (1 - PE) \times m_{dry\ grounds}$. Substitution into again yields equation (6.5).

Hopefully the expression in equation (6.5) is intuitive: the higher the TDS in the brew, the higher the extraction from the solid phase must have been. The important point is that measurement of the brew TDS, along with weighing the dry grounds and the brew, gives an estimate of the percent extraction – and a useful measure of whether you are under- or over-extracting your coffee.

Part A – Exploration of Mass Transfer & Extraction

Our primary experiments today will focus on comparing different mass transfer or extraction protocols to assess their effects on the final taste and amount of "coffee stuff" extracted into the brew. We will use a new brewing technique, the French press, to set up several extraction tests. For each brew we will measure the TDS in the brewed coffee with a digital refractometer, calculate the corresponding PE, and perform sensory evaluations.

1) First, measure the TDS of plain tap water. A normal reading will be somewhere around 0.01% to about 0.05% (compared to coffee, which is around 1.25%). Repeat at least two more times during the lab session. One reason to do this is to get a standard deviation, i.e., a measure of reproducibility. Another reason is to make sure that the refractometer is still calibrated properly and that the sample well is not dirty.

PROPER USAGE OF THE GRINDERS

- If using a Baratza grinder, do NOT use the extreme values. Too coarse (near 40) can cause the lid to pop off. Too fine (near 1) can clog the grinder internally.
- For Baratza grinders, only use grind settings between 5 and 35.

PROPER USAGE OF THE REFRACTOMETERS

- **Let the coffee cool!** The meters will give error messages at high temperature.
- To cool it quickly, pipette a little bit into a cup, let it cool off for a few minutes, then pipette just a bit into the sample well of the refractometer.
- Only pipette enough to cover the glass sensor - don't add too much.
- Don't put any solid coffee grounds on the refractometer, only liquid.
- Don't push the "menu" button... you don't need to change any settings.
- After you're done **use only a Kimwipe** (not a paper towel) to clean the sensor.
- Don't move the refractometers... please be patient and wait for your turn.

2) Next, perform three simultaneous brews in the French presses to assess the impact of **Surface Area**. Use the same brew ratio, water temperature (~94°C), and extraction time (5 minutes), and compare 3 different grind sizes:

   i. Coarsely ground (35 on a Baratza, or pestle and mortar if you like)
   ii. Medium ground (a 'normal' grind, around 18 on a Baratza)
   iii. Finely ground (around 5 on a Baratza)

   After the extraction time use the metal press to push the coffee grounds down, then pour out a small sample into a cup. Use a plastic pipette to draw out a small sample from the cup and place the brewed coffee into another cup to cool it quickly before measuring in the refractometer. Make sure you taste each brew, record your sensory impressions, and measure the TDS as well as the mass of the brewed coffee (which you get from the difference in mass of the full cup and the empty cup). You need <u>both</u> the brew mass and TDS to calculate the PE.

3) The next goal is to assess the impact of **Temperature.** Perform two more brews, using the same brew ratio, same 5-min extraction time, and same medium grind. Compare two temperatures:

   i. Moderate temperature (~70 °C)
   ii. Very hot temperature (around 99 °C).

   Again, make sure you taste and record your sensory impressions, and measure both the mass of the final brew and the TDS. Be careful when you taste the very hot one!

4) The final goal is to quantify the effect of **Extraction Time**. Prepare one more brew, again using the same brew ratio, medium grind size, and ~94°C temperature. But this time, leave the metal press off – you are going to sample from the middle of the liquid every 60 seconds for 10 minutes. Follow this procedure:

   i. Put the coffee grounds in the press (leave the metal press aside).
   ii. Pour in the hot water, making sure to wet all the grounds, and as soon as done pouring start a timer.
   iii. Immediately after pouring is complete, extract a small sample from the center of the liquid using a pipette. Put it into another cup to let it cool.
   iv. Every 60 seconds, use the pipette to extract another small sample from the center of the liquid. Repeat until 10 minutes.

   At the end of the 10-minute brew, put the metal press on, push it down, and then weigh the mass of the brew and taste. Don't worry about small mass loss from sampling, you can neglect that and use the brew mass to estimate the extraction yield.

TDS Data for Impact of Surface Area

Coffee type____

Mass of hot water: _______ *grams* Mass of grounds:________ *grams* R_{brew}:________

Temperature of hot water: _______ °C Extraction time: ______ *minutes*

Grind size: COARSE

Mass of empty cup: _______ *grams* Mass of filled cup: _______ *grams*

Mass of brew: _______ − _______ = ________ *grams*

TDS: ________ % PE: ________ × ________ ÷ ________ = ________ %

TDS × m_{brew} ÷ $m_{grounds}$

Sensory Evaluations:

____

____

Grind size: MEDIUM

Mass of empty cup: _______ *grams* Mass of filled cup: _______ *grams*

Mass of brew: ________ − ________ = ________ *grams*

TDS: ________ % PE: ________ × ________ ÷ ________ = ________ %

Sensory Evaluations:

____

____

Grind size: FINE

Mass of empty cup: _______ *grams* Mass of filled cup: _______ *grams*

Mass of brew: ________ − ________ = ________ *grams*

TDS: ________ % PE: ________ × ________ ÷ ________ = ________ %

Sensory Evaluations:

____

____

TDS Data for Plain Water

1st measurement: ______ % 2nd measurement: ______ % 3rd measurement: ______ %

TDS Data for Impact of Temperature

Coffee type____

Mass of hot water: _______ *grams* Mass of grounds: _______ *grams* R_{brew}:________

Grind type: ____________ Extraction time: ______ *minutes*

Temperature: <u>MODERATE TEMPERATURE</u>

Temperature of water: _______ °C

Mass of empty cup: _______ *grams* Mass of filled cup: _______ *grams*

Mass of brew: ________ – ________ = ________ *grams*

TDS: ________ % PE: ________ × ________ ÷ ________ = ________ %

Sensory Evaluations:

____

____

____

Temperature: <u>VERY HOT</u>

Temperature of hot water: _______ °C

Mass of empty cup: _______ *grams* Mass of filled cup: _______ *grams*

Mass of brew: ________ – ________ = ________ *grams*

TDS: ________ % PE: ________ × ________ ÷ ________ = ________ %

Sensory Evaluations:

____

____

____

TDS Data for Impact of Extraction Time

Coffee type____

Mass of hot water: _______ *grams* Mass of grounds: ______ *grams* R_{brew}:_________

Grind type: ____________ Temperature of hot water: _______ °C

Extraction Time: <u>10 Minutes, Measured every Minute</u>

Mass of empty cup: _______ *grams* Mass of filled cup: _______ *grams*

Mass of brew: ________ – ________ = ________ *grams*

Time (minutes)	TDS	Time (minutes)	TDS
____________	______________	_____________	______________
____________	______________	_____________	______________
____________	______________	_____________	______________
____________	______________	_____________	______________
____________	______________	_____________	______________

Sensory Evaluations: ____

____

Roasting Data

1st Roast Coffee type: _______________________ Roaster settings: ____________

Mass of green beans: ________ *grams* Mass of roasted beans: ________ *grams*

First crack: ______ *mins* Roast time: ______ *mins* Color: ________________

Energy usage: ________ *kW-hr* Notes:_____________________________________

____

2nd Roast Coffee type: _______________________ Roaster settings: ____________

Mass of green beans: ________ *grams* Mass of roasted beans: ________ *grams*

First crack: ______ *mins* Roast time: ______ *mins* Color: ________________

Energy usage: ________ *kW-hr* Notes:_____________________________________

____

Part B – Roasting

Do two roasts for your next lab session. Try different beans and roast levels to try to optimize the quality. Coffee beans vary wildly, so you will need to optimize your roast around your particular beans. Feel free to experiment with your fan and power settings, and make sure you record those settings, the mass of green and roasted beans, the final color, and most importantly your energy usage in kW-hr!

Lab Report

By the due date, each group will submit their lab report that includes: (1) a labeled column plot that shows the TDS value for each of your French press brews, (2) a labeled column plot that shows the PE for each of your brews, (3) a scatter plot of TDS vs. extraction time for your 10-minute brew, and (4) a paragraph answering the questions below.

(1) Enter your TDS data into Excel or Google Sheets and prepare a "column plot" of TDS versus the different brewing conditions. Each column should be properly labeled with all three brewing parameters (e.g., "Coarse, 94 °C, 5-min"). Use "insert text" to put some abbreviated tasting notes next to each column. You should have a total of six brews (use the final TDS from the 10-min brew).

(2) Using equation (6.5), calculate the PE for each of the brews. Prepare another column plot of the PE versus brewing conditions and again make sure each column is properly labeled.

(3) Next, prepare a scatter plot of the TDS versus time for your 10-minute brew. Does it follow the expected trend?

(4) Finally, on a separate page answer the following questions:

   i. What is the TDS of tap water? Do you think the number you obtained is reasonable? Why or why not? What is the TDS in PPM? (See the Lab 6 bonus box.)

   ii. Based on your data, which brewing parameter had the most impact on the final TDS: grind size, temperature, or extraction time? Why do you think this was the case?

   iii. Did you see any differences in trends with PE when compared to the trends in TDS? If so, what might account for these differences?

   iv. Recall the ideal extraction is typically considered to be about 20%, with a TDS near 1.3%. Which of your brewing conditions yielded a brew closest to those ideals? Did you think this was the best tasting coffee you brewed? How did it taste compared to the others? (Remember that the best measuring device for coffee taste isn't the TDS meter… it's your taste buds!) What is the TDS of your best brew in PPM?

   v. Describe how the TDS varied with extraction time. Why do you think it behaved like this? What do you think would have happened to the TDS and PE if you let it brew for 20 minutes?

Lab 6 Bonus Box – What the Heck is "ppm?"

Liquid	Approximate Dissolved solids
Distilled water	1 ppm
Bottled water	250 ppm
"Hard" tap water	450 ppm
Brackish water	3,000 ppm
Black coffee	13,000 ppm
Sea water	35,000 ppm

When the concentration of something is really, really small, one of the units that might be used is "parts per million," or ppm. For example, the amounts of certain naturally occurring chemicals in drinking water that are desirable (like sodium, magnesium, fluoride) or undesirable (like lead and arsenic) are often listed in ppm. Most cities issue water quality reports if you are interested in what your tap water contains. If you use tap water for your coffee brewing, then ultimately your coffee also has those chemicals! For example, in Davis, California the average level of sodium is 85 ppm, chloride is 21 ppm, and fluoride is 0.3 ppm. Some levels are so low that parts per billion (ppb) are used, like for lead which is less than 5 ppb.

So, what exactly does ppm mean? If we use fluoride (F) as an example, 0.3 ppm means that there are 0.3 grams of fluoride per 1,000,000 grams of tap water. The units of measure here are grams, but any unit of measure can be used as long as they are the same for both quantities – like pounds or kilograms.

$$1 \text{ ppm F} = \frac{1 \text{ gram of F}}{1{,}000{,}000 \text{ grams of water}}, \text{ or in general, } 1 \text{ ppm} = \frac{1 \text{ unit of stuff}}{1{,}000{,}000 \text{ units of water}}$$

Fluoride is interesting because unlike Davis, most cities and towns supplement their water with fluoride to reach the level of 0.7 ppm recommended by the federal government's Health and Human Services Department. This level of fluoride helps prevent dental tooth decay. Too much fluoride in your water, however, results in dental fluorosis – a permanent change in the appearance of the tooth enamel, varying from white spots in mild forms to staining or even pitting in severe cases. Fortunately, dental fluorosis is only really a concern when your teeth are developing, which is why children are taught to spit out toothpaste and rinse after brushing.

But what do ppm and different chemicals in water have to do with coffee? Well, coffee contains a cornucopia of different chemical molecules extracted from the roasted coffee beans in the ppm level. When you are measuring TDS to quantify all the stuff in the brewed coffee, the ideal is around 1.3% TDS, which if you convert to ppm is

$$1.3\% \text{ TDS} = \frac{1.3 \text{ grams of solids}}{100 \text{ grams of water}} = \frac{1.3 \text{ grams of solids}}{100 \text{ grams of water}} \times \left(\frac{10^6 \text{ grams of water}}{10^6 \text{ grams of water}}\right)$$

$$= \frac{(1.3\times10^6 / 100) \text{ grams of solids}}{10^6 \text{ grams of water}} = \frac{13{,}000 \text{ grams of solids}}{10^6 \text{ grams of water}} = 13{,}000 \text{ ppm.}$$

That might sound like a lot, but many of the individual and important flavor molecules are in the 10 to 100 ppm range or lower (remember there are about 1000 different molecules that contribute to the flavor of coffee!) The low levels of all sorts of different chemicals in the coffee is part of what makes it so darn difficult to optimize brewed coffee through chemical analysis. The myriad of chemical reactions that occur during roasting are too complex to fully analyze, and the low concentrations make it hard to quantify their effect on the brew's taste profile. Moreover, people's sensitivity to the various flavor molecules varies. Add the fact that the chemical content of the starting water is also variable based on the water source, and the analysis and optimization of coffee becomes a very difficult scientific problem. Of course, you should use all the analytical tools you have available, but at the end of the day, the best measuring devices are your own taste buds. We will explore water quality next in Lab 7.

Lab 7 – Pressure Driven Flow through Coffee Grounds

Objectives: In this lab we will study "fluid mechanics" as applied to the motion of water through coffee grounds. We will relate the concepts of pressure and flow rate using an equation known as "Darcy's Law," and we will assess how the flow rate affects the strength and quality of the brewed coffee.

Equipment:

☐ AeroPress ☐ Electric kettle ☐ Digital refractometer ☐ Bathroom scale

☐ Ruler ☐ Kill-a-Watt meter ☐ Roaster

Activities:

☐ Part A – four brews with the AeroPress, to examine Darcy's law & pressure

- ☐ **Review safety guidelines!** Then a practice brew to gauge your applied pressure
- ☐ Three brews at different applied pressures

☐ Part B – four more brews with the AeroPress, to examine permeability

- ☐ Two brews at same pressure but different grind sizes
- ☐ Two brews at same pressure and grind, but different masses of grounds

☐ Part C – two or more roasts to have beans for the next lab

Report:

☐ Scatter plot of flow rate versus the applied pressure gradient

☐ Scatter plot of TDS versus flow rate

☐ Paragraph discussing tasting notes and key questions about the lab

Background – Fluid Mechanics

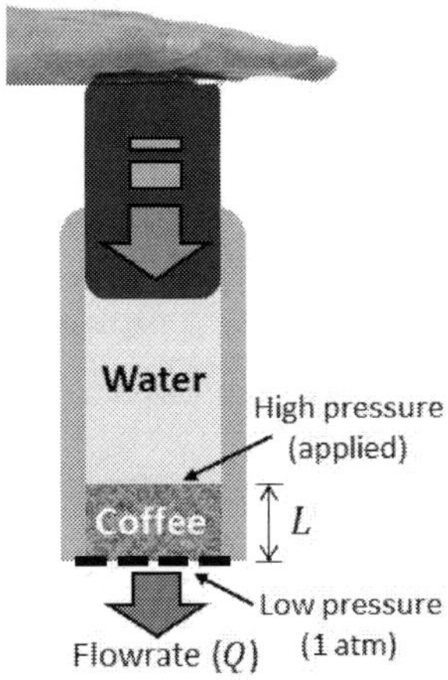

By this point, you should understand that the "extraction time" is a critical parameter that affects the final taste of the coffee: too short of a time yields sour and "under-extracted" coffee, but too long of a time yields bitter and "over-extracted" coffee. In most methods of brewing coffee, the extraction time is controlled by how quickly the hot water moves past the coffee grounds. In other words, the fluid velocity is a key parameter, so understanding 'fluid mechanics' is necessary.

As a reminder, the motion of fluids, like all other forms of matter, is governed by Newton's second law, i.e., $F = ma$ (where F is the sum of the forces acting on the fluid, m is the mass, and a is the acceleration). Unlike the rigid objects (like cannonballs etc.) you might have studied previously in physics courses, a key difference for fluids is that they are deformable.

This difference tremendously complicates interpretation of the forces acting on fluids, so there are entire courses in engineering curricula dedicated just to fluid mechanics.

For this lab, we will focus on how liquids move through "porous media," in this case coffee grounds. Specifically, we use Darcy's Law, which states that the velocity u of the liquid (in cm/s) is proportional to the pressure difference across the porous medium:

$$u = -\frac{\kappa}{\mu}\nabla P = \frac{\kappa}{\mu}\left(\frac{P_{entrance}-P_{exit}}{L}\right). \tag{7.1}$$

There are several terms here discussed earlier in this chapter; here we briefly review them. The "viscosity" μ of the fluid characterizes how easy it is for it to deform or to "shear." Water has a low viscosity, whereas honey has a very high viscosity. For the purpose of this lab, we will approximate the viscosity of the coffee with that of water (since the coffee is 99% water). The "permeability" κ characterizes how much open space there is in the porous medium versus solid surface area: the higher the surface area to volume ratio, the more surface area is available to "slow" down the fluid. As a result, finely ground coffee has a higher surface-area-to-volume ratio and smaller permeability than coarsely ground coffee. Recall that the surface-area-to-volume ratio also affects mass transfer during extraction: there is a much larger surface area with smaller particle sizes so the rate of mass transfer is much faster.

Finally, the pressure gradient $\nabla P \approx \Delta P/L$ is the "driving force" for the fluid flow. (Recall that pressure is force/area.) Note that the "gradient" in pressure is how much the pressure changes over the length L of the porous medium. When you use the Mr. Coffee or a pour-over technique, the pressure difference is provided by gravity acting on the weight of the water. Other devices, however, provide more control over the pressure gradient (most famously in espresso machines).

The applied pressure P_{top} is the pressure you apply by hand plus the atmospheric pressure, while the pressure P_{bot} is the pressure in the air outside the bottom of the AeroPress, which is just the atmospheric pressure. In other words, the pressure difference is

$$\Delta P = P_{top} - P_{bot} = \left(P_{applied} + P_{atm}\right) - P_{atm} = P_{applied}. \tag{7.2}$$

Note the atmospheric pressure cancels out because it pushes on both the top and bottom.

How do we measure the applied pressure? In this lab, we will use a standard bathroom scale (what you typically use to measure your body weight). When you put the AeroPress on the scale and push on it, you can see how many pounds of force you apply by reading the scale. The force will naturally fluctuate a little bit, but with some practice you can keep the applied force approximately constant during the AeroPress brew. Say you apply 20 pounds of force. The corresponding pressure is this force divided by the cross-sectional area of the plunger in the AeroPress, which is about 1.6 inches in radius. The area is $A = \pi r^2$. Thus,

$$P_{applied} = \frac{F_{applied}}{Area} = \frac{20\ lbf}{3.14 \times (1.6\ inch)^2} = 2.5\ \frac{lbf}{in^2} = 2.5\text{ psi}. \tag{7.3}$$

Here lbf means "pounds force" and psi stands for "pounds per square inch." One "atmosphere" of pressure is 14.7 psi. If you applied 200 pounds of force (don't try this!) you would have 1.7 atmospheres of pressure. Espresso machines apply about 9 atmospheres of pressure. To finish the example, if you then measure the thickness L of the coffee grounds as 0.5 inches, then your pressure gradient was

$$\nabla P = \frac{P_{applied}}{L} = \frac{2.5\ psi}{0.5\ inch} = 5\ \frac{lbf}{in^3} \tag{7.3}$$

A key goal for this lab is to gain a semi-quantitative appreciation for Darcy's law and how quickly fluid flows through the porous coffee grounds and affects the taste of the resulting cup. The lower the applied pressure, the longer the effective extraction time will be – with a corresponding impact on the mass transfer and ultimate sensory qualities of the brew. Likewise, you can use a much finer grind size and therefore a shorter brewing time. Which approach is better? In this lab you get to test for yourself!

DANGER OF BURNS! PROPER USE OF THE AEROPRESS

Be very careful with the AeroPress – it is easy to burn yourself if careless!

- The maximum capacity is about 200 grams of water.
- **DO NOT ADJUST THE AEROPRESS AFTER YOU PUT HOT WATER IN IT.** You likely will spill hot water and burn yourself.
- Before dispensing, **double-check that the filter cap is hand-tight** and secure. If it is loose, the brew will fall out onto your hand and burn you badly when you invert it. If it is crooked, brew will leak out the side.
- Hold the AeroPress steady with one hand as you apply an evenly applied pressure straight down. Avoid pushing it sideways or at an angle.
- Don't forget the filter paper – otherwise the brew will rush out, and you'll get a gritty mess.

Part A – Pressure driven flow in the AeroPress

We will perform eight brews using the AeroPress to get a feel for Darcy's law. First, we will review the safety guidelines, then do a practice brew (without any measurements) to get a sense for how the brewer works. Then, we will systematically measure the flow rate and corresponding TDS for different applied pressures, different grind sizes (which changes the permeability κ), and different masses of coffee grounds (which alters the thickness L). In each case we will measure the time required for the water to move through the grounds.

1) Before doing anything, review the safety guidelines for the AeroPress in the box above.
2) To brew with the AeroPress, follow this procedure.
   i. Position the plunger at line "4" if there are markings. If there are no markings, adjust the plunger so that it is fully seated at the bottom of the reservoir with all the rubber inside, with the hexagonal rim on the opposite, open side.
   ii. Place the AeroPress on your mass balance on a counter or sturdy table so that the plunger is down, and the open cylinder portion with the hexagon is up.
   iii. Add your ground coffee. Then, carefully add a measured mass of hot water, making sure to wet all the grounds and stir the mixture briefly. The AeroPress has a small volume: for brew ratio 15, add 12 grams of grounds and 180 grams of water.
   iv. Place a filter in the cap. Secure the cap on top snugly, then wait for the desired amount of extraction time.
   v. Finally, doublecheck that the filter cap is on securely, then quickly and carefully invert the AeroPress onto a large glass mug or measuring cup, ideally on a bathroom scale placed on a counter so that you can measure the applied force.
   vi. Carefully press down until you finish pushing out the brewed coffee (you will feel it change when it starts pushing air through the coffee grounds). Keep an eye on the bathroom scale to assess how much force you are applying.

3) We will perform the first four trials all at the same grind size. Prepare enough ground coffee for the four brews in the AeroPress (about 50 grams). Heat water in the kettle.
4) For the first practice brew, don't worry about using the bathroom scale. Put 12 grams of coffee grounds inside the AeroPress, and then add hot water. Give a quick stir and wait 4 minutes to provide some extraction time. Then, invert the AeroPress over a ceramic or glass cup on the bathroom scale and apply a "gentle" pressure. Do your best to maintain the same force while dispensing.
5) Taste the brew. What are your sensory impressions?
6) For the second trial, repeat the process with everything identical, but this time be prepared to take more measurements. Place the cup and AeroPress on the bathroom scale and ready a cell phone timer to measure how much time it takes to dispense the brew. When dispensing, do your best apply a "gentle" continuous pressure (e.g., 10 lbf), and record the average pounds of force on the bathroom scale. Record the time $t_{dispense}$ required to dispense the fluid, which is the time from when you first start pushing until you feel it change to pushing air through the compacted coffee grounds. (Don't discard the spent grounds yet until after step 7!)
7) Weigh the mass m of brewed coffee in the cup. If we approximate the density of brewed coffee as $\rho = 1$ gram / cm^3 (close to that of water), then the volume delivered is estimated as $V = m/\rho$. The average flowrate is then calculated as $Q = V/\, t_{dispense}$. Calculate and record this flowrate.
8) Carefully uncap the AeroPress and fully extend the plunger while holding it vertically. Use a ruler to measure the thickness L of the "puck" of spent coffee grounds. Then measure the TDS and taste the brew. What are your sensory impressions? Discard the puck, rinse and prepare for the next brew.
9) Repeat this process for two more brews but increasing the applied force each time: first a "moderate" force (e.g., 20 lbf), and then a "strong" force (e.g., 30 lbf). There is no prize for being too strong! If you apply too much force, the filter paper can tear or fluid can leak out the side of the filter cap. Each time, record the dispensing time, the mass of fluid delivered and corresponding flow rate, the applied force, the thickness L of the spent coffee grounds, and the resulting TDS and sensory qualities.

Part B – Permeability and path length in the AeroPress

In this part, our procedure is almost exactly the same as in Part A, except here we will focus on the effect of permeability and path length via the grind size and total mass of grounds.

1) Next, compare the effect of grind size. Grind 12 grams of coffee as fine as possible, and another 12 grams as coarse as possible. Perform an AeroPress brews with the fine grind, using the same "moderate" force to the best of your ability, then repeat with coarse grind, again with the same force. For both brews record all measurements.
2) Finally, compare the effect of the path length L. Grind 30 grams of coffee to medium grind size. Repeat two final AeroPress brews, one using just 6 grams of coffee, and the other using 24 grams. Using the same "moderate" force to the best of your ability, and record all measurements.

Data for Pressure Experiments

Keep the following parameters the same for all brews here unless otherwise specified.

Coffee type____

Mass of hot water: _______ *grams* Mass of grounds: _______ *grams* R_{brew}: _______

Temperature of hot water: _______ °C Grind size: ___________

Initial extraction time: ______ *minutes*

<u>Gentle Pressure</u>

Mass of brew: ________ – ________ = ________ *grams* (filled cup – empty cup)

Applied force: ________ *pounds* Thickness of spent coffee grounds: _________ *cm*

Total dispensing time with pressure applied: ______ *seconds*

Flowrate: ________ cm^3 ÷ ________ *seconds* = ________ cm^3 / *second*

TDS: ________ % PE: ________ × ________ ÷ ________ = ________ %

Sensory Evaluations: ____

____

<u>Medium Pressure</u>

Mass of brew: ________ – ________ = ________ *grams* (filled cup – empty cup)

Applied force: ________ *pounds* Thickness of spent coffee grounds: _________ *cm*

Total dispensing time with pressure applied: ______ *seconds*

Flowrate: ________ cm^3 ÷ ________ *seconds* = ________ cm^3 / *second*

TDS: ________ % PE: ________ × ________ ÷ ________ = ________ %

Sensory Evaluations: ____

____

<u>Strong Pressure</u>

Mass of brew: ________ – ________ = ________ *grams* (filled cup – empty cup)

Applied force: ________ *pounds* Thickness of spent coffee grounds: _________ *cm*

Total dispensing time with pressure applied: ______ *seconds*

Flowrate: ________ cm^3 ÷ ________ *seconds* = ________ cm^3 / *second*

TDS: ________ % PE: ________ × ________ ÷ ________ = ________ %

Sensory Evaluations: ____

____

Data for Grind Size Experiments

Moderate Pressure, Coarse Grind

Grind size: ________

Mass of brew: ________ – ________ = ________ *grams* (filled cup – empty cup)

Applied force: ________ *pounds* Thickness of spent coffee grounds: ________ *cm*

Total dispensing time with pressure applied: ______ *seconds*

Flowrate: ________ cm^3 ÷ ________ *seconds* = ________ cm^3 / *second*

TDS: ________ % PE: ________ × ________ ÷ ________ = ________ %

Sensory Evaluations: ____

____

Moderate Pressure, Fine Grind

Grind size: ________

Mass of brew: ________ – ________ = ________ *grams* (filled cup – empty cup)

Applied force: ________ *pounds* Thickness of spent coffee grounds: ________ *cm*

Total dispensing time with pressure applied: ______ *seconds*

Flowrate: ________ cm^3 ÷ ________ *seconds* = ________ cm^3 / *second*

TDS: ________ % PE: ________ × ________ ÷ ________ = ________ %

Sensory Evaluations: ____

____

Permeability calculation: Your data collected here allows you to calculate the permeability, in units of cm^2. To do this, solve Darcy's law for the permeability κ, then insert your experimental values for Q and $P_{applied}/L$. Estimate the viscosity as simply that of water near 80°C, which is $\mu \approx 0.35 \times 10^{-3}$ Pa·s. Here Pa·s is read as "Pascal second," where the Pascal is a unit of pressure equal to 1 Newton per square meter. There are 6895 Pascals in 1 psi. How much does your calculated permeability vary with grind size?

Data for Coffee Mass / Path Length Experiments

Moderate Pressure, Small Mass Mass of grounds: _______ *grams*

Mass of brew: ________ – ________ = ________ *grams* (filled cup – empty cup)

Applied force: ________ *pounds* Thickness of spent coffee grounds: _________ *cm*

Total dispensing time with pressure applied: ______ *seconds*

Flowrate: ________ cm^3 ÷ ________ *seconds* = ________ cm^3 / *second*

TDS: ________ *%* PE: ________ × ________ ÷ ________ = ________ *%*

Sensory Evaluations:____

Moderate Pressure, Large Mass Mass of grounds: _______ *grams*

Mass of brew: ________ – ________ = ________ *grams* (filled cup – empty cup)

Applied force: ________ *pounds* Thickness of spent coffee grounds: _________ *cm*

Total dispensing time with pressure applied: ______ *seconds*

Flowrate: ________ cm^3 ÷ ________ *seconds* = ________ cm^3 / *second*

TDS: ________ *%* PE: ________ × ________ ÷ ________ = ________ *%*

Sensory Evaluations:____

Part C – Height versus time in a gravity-drained Aeropress

Next, we will move away from applying our own pressure gradient, and instead let them drain entirely by gravity. The goal is to see how well the predictive model from this chapter describes the behavior of this more complicated experimental system.

1) Place the filter paper and cap on the Aeropress; set the plunger to the side (you won't need it). Put the Aeropress, filter down, on a cup, and put in about 15 grams of finely ground coffee.
2) Have one groupmate hold a ruler up near the side of the Aeropress, so that you can measure the liquid height level versus time. Have another groupmate ready to record a video with their phone. Have another person make a written record of the height vs. time (in case the video doesn't work for any reason).
3) To begin, carefully pour 200 grams of hot water into the Aeropress and begin recording. Let the fluid drain by gravity. Continue recording until the height of the fluid is at the top of the bed of coffee grounds.
4) After it's done, carefully insert the plunger, remove the filter cap, and push out the filter bed to measure its thickness.
5) Repeat the experiment with a different grind size.

Part D - Roasting

Finally, do two or more roasts for next time. Keep track of your masses and energy usage! The design competition is fast approaching, so optimizing your roasting will be a critical aspect. Have you found a roast profile for the available green beans that you like?

Data for Gravity-Drained Aeropress

Mass of grounds: _______ *grams* Thickness of spent coffee grounds: _________ *cm*

Mass of brew: ________ – ________ = ________ *grams* (filled cup – empty cup)

Time (sec)	**Height (cm)**	**Time (sec)**	**Height (cm)**	**Time (sec)**	**Height (cm)**
______	_______	______	_______	______	_______
______	_______	______	_______	______	_______
______	_______	______	_______	______	_______
______	_______	______	_______	______	_______
______	_______	______	_______	______	_______
______	_______	______	_______	______	_______
______	_______	______	_______	______	_______
______	_______	______	_______	______	_______
______	_______	______	_______	______	_______
______	_______	______	_______	______	_______
______	_______	______	_______	______	_______

Roasting Data

1st Roast Coffee type: _______________________ Roaster settings: ______________

Mass of green beans: ________ *grams* Mass of roasted beans: ________ *grams*

First crack: ______ *mins* Roast time: ______ *mins* Color: ________________

Energy usage: ________ *kW-hr* Notes:_____________________________________

____

2nd Roast Coffee type: _______________________ Roaster settings: ______________

Mass of green beans: ________ *grams* Mass of roasted beans: ________ *grams*

First crack: ______ *mins* Roast time: ______ *mins* Color: ________________

Energy usage: ________ *kW-hr* Notes:_____________________________________

____

Lab Report

By your specified due date, each group will submit their lab report that includes (1) a scatter plot of the fluid velocity versus the pressure gradient, (2) a summary of calculations for the fluid velocity, (3) a scatter plot of the TDS versus fluid velocity, (4) scatter plots of the liquid height versus time in both linear and logarithmic scale, and (5) a paragraph answering the questions below.

(1) Prepare a scatter plot that shows the average fluid velocity versus the applied pressure gradient, $\Delta P/L$. There should be at least three data points from part A; use the same color and shape marker to indicate the grind size. There should be another 3 data points from Part B; use different colors to indicate how these differed. Make sure you label your axes! The plot should include a linear regression for all the data points corresponding to the same permeability (which ones should these be?)

(2) On a separate page clearly show how you calculated velocity from your raw data, as well as a description of how you extracted the effective permeability from your scatter plot. What is the value of the permeability? What did you need to assume about the viscosity?

(3) Prepare a scatter plot of the TDS versus the velocity. Does the TDS increase or decrease with the average fluid velocity?

(4) Prepare a scatter plot of the liquid height versus time for the gravity draining experiment, for both grind sizes. Prepare a second version of this plot that shows h/h_0 on a logarithmic scale (i.e., the vertical scale is logarithmic). What trends do you observe? What might explain any deviations based on what you expect from Darcy's Law?

(5) Using the logarithmic plot, devise a procedure to estimate the permeability from the gravity draining experiments – how does the permeability here compare to the ones you calculated from the applied pressure experiments? Clearly show your calculations.

(6) Finally, write a brief paragraph that addresses the following questions:

(i) How accurately did Darcy's law describe the flow rate through the coffee grounds when you applied pressure by hand? Did you obtain a linear trend between Q and $P_{applied}/L$ for your medium grind?

(ii) How did the increased amount of grounds affect the rate at which fluid moved through the grounds in the AeroPress? Did it affect the magnitude of *L?* Was it linear? How did the grind size affect the velocity?

(iii) How did the TDS vary with the flowrate? What trends did you observe? In general, does the flowrate serve as a good predictor for what the TDS will be?

(iv) How well did the draining column analysis predict the behavior of the gravity-drained Aeropress? Was there an exponential decay in the height? If not, what physical effects might account for any differences between the model(s)?

(v) How did the different flow rates and conditions affect the taste of the coffee?

Lab 8 Bonus Box – What Makes Espresso Different?

If you walk into a modern coffeehouse, you are likely to see a long menu of different coffee based beverages. Lattes, cappuccinos, macchiatos, mochas, flat whites, Americanos, and red eyes are all typically available even at the smallest hole-in-the-wall café. Although these drinks feature different amounts of milk or other ingredients, they all share something in common: specifically, they all feature "espresso" as a central ingredient.

What is espresso? Some people mistakenly think of espresso as a different type of roast, but espresso is really a brewing method. In fact, espresso is very similar to brewing coffee with an AeroPress, except with *very* high applied pressures. With your hand on the AeroPress, you're probably able to apply up to around 60 pounds of force, equivalent to about 7 psi (or 0.5 atmospheres). In contrast, the hot water inside an espresso machine is typically pressurized to around 130 psi, which is equivalent to about 9 atmospheres!

How does the espresso machine generate such a high pressure? Espresso was invented in Italy in the early 1900s, and many of these earliest espresso machines actually had a large lever that the barista would pull downward by hand to pressurize the water. This is why you often hear people talk about "pulling a shot" of espresso, even though modern baristas don't actually pull on anything. (Those early hand-lever espresso machines were quite dangerous: if your hand slipped, the lever would fly back up and potentially break your jaw!) Instead of a lever, many modern espresso machines use the "boiler," an enclosed chamber inside the machine, to boil the water and pressurize the steam up to 9 atmospheres. More expensive machines feature "positive displacement pumps" that provide precise control of the pressure and use water that is typically heated between 91 to 94 °C.

With all this pressure available, very high flow rates through the coffee grounds become possible. In fact, the name "espresso" alludes to the idea that this is a fast or "express" brewing technique. When you watch a shot being pulled, though, it might look like it's coming out pretty slowly, filling up only a small espresso cup in 30 or so seconds. Just like the AeroPress, the flow rate for espresso is governed by Darcy's law. Unlike the AeroPress, the coffee beans for espresso are typically ground extremely fine, which maximizes the surface area to volume ratio to maximize the rate of mass transfer (think back to Lab 6). At the same time, though, the very small grind size for espresso also decreases the permeability κ significantly. As a result, a typical shot of espresso has a small flowrate despite the huge applied pressure.

The role of the permeability helps give us insight on why baristas must practice their technique to pull a good shot. Minor changes in the grind size have a big impact on the permeability, and hence the flow rate and corresponding mass transfer – resulting in a big change in the flavor. Baristas spend a lot of time "dialing in" their grind size. Furthermore, they need to perfect their "tamping" technique, where they use a heavy metal tamp to squish the dry coffee grounds into a well-packed and level puck inside the "portafilter" that gets clamped into the espresso machine. It is not easy to get the right mass and thickness every time!

All this effort is worth it, however, because a good shot of espresso will be sweet, syrupy, and have a beautiful "crema" of foam on top resulting from the carbon dioxide released rapidly from the coffee beans. An espresso by itself should be drunk as soon as possible so that you can enjoy the crema before it dissipates. Of course, if you prefer milky drinks like lattes or cappuccinos, then the highly concentrated shot of espresso adds the desired intense coffee flavor. Whichever you prefer, thanks Darcy's law for making it possible!

Part III

Engineering Design

Design Competition Format, Guidelines, & Video Project

The first seven labs of this book have focused on "analysis" to help you understand the various scientific principles that relate to the quality of coffee. We now shift our attention to "engineering design," where our goal is to design a coffee that satisfies specified criteria. In this section, we describe the format for a coffee design competition and video project that has been extremely popular at UC Davis. It's now time to think carefully about the engineering design goals, since they will help you think quantitatively about how to design a process – a skill that chemical engineers apply to other design goals (for example, making the most efficient biofuel for the lowest price).

Here, the main goal of the coffee design project is to make the **best** cup of coffee, using the **least** amount of electrical energy. Each group will receive a score defined by the ratio

$$\text{Final Score} = \frac{\text{Blind Taste Test Score}}{\text{Total Electrical Energy}}$$

Note carefully what this metric implies: it is very possible for you to make the best tasting coffee, but to still lose the design competition because your process used too much energy!

If there are sufficient people competing, then there are two rounds of competition. In the first round (the "playoffs"), groups will compete within their section. The group in each section that receives the highest final score will then advance to the second round (the "championship") to compete for the title of grand champion. The two inputs into the final score are defined as follows.

- The coffee will be judged in a blind taste test on a scale of negative 5 to positive 55 (where –5 is amazingly bad coffee, and +55 is stunningly good.) The point distribution is loosely based on the official tasting guidelines established by the Specialty Coffee Association (cf. Lab 1). All of the blind taste scores for each coffee will be averaged together.

- Your group must keep track of how many kilowatt-hours of electrical energy you used during the entire process of roasting and brewing your coffee. Your roasting energy will be "normalized" by how much mass of any particular roast you used brewing. That means it is OK to roast more than you need for the competition… you only "pay" for the energy to roast the beans you actually use.

- Your brewing energy, however, will not be normalized. If you heat up too much water… then you have to account for all of it! Any piece of equipment you plug into an outlet must be monitored using the Kill-a-Watt meter and included in your final tally. This is typically only the brewer or kettle and the roaster; as we saw in Lab 5 the energy of the grinder is negligible so we won't specifically keep track of it here.

Groups who place in the top three in either taste score or taste per energy score will receive some bonus points, but only the winning group in terms of taste per energy will advance to the championship round. The group with the highest taste per energy score in the championship round will receive additional bonus points, as well as a grand prize. If there is a championship round, the bonus points can be doubled for placing in the top three for the overall championship.

The following design constraints must be followed.

1) **Ingredients:** The only ingredients you are allowed to use are green coffee beans and water. No syrups, sweeteners, spices, or other artificial flavors may be added.

2) **Minimum Volume:** On the day of the contest, we will use a random number generator between 500 to 925 to determine the amount of coffee, in grams, required for the competition. This requires the competitors to be able to rapidly scale their designed "best tasting coffee while minimizing energy consumption" in real time for the competition.

3) **Energy Penalty:** If you don't deliver the minimum amount of required brew, you will be assessed an energy penalty of 0.005 kW-hr per gram of missing brew. (If you make more than the required amount, the energy penalty is zero.)

4) **Time Allowed:** On the day of the blind tasting, you will have 45 minutes to brew your approximately 1 liter of coffee. Brew too early and your coffee may lose some of its freshness, brew too late and risk being disqualified!

5) **Equipment:** You are free to use any combination of the equipment available in the lab for roasting and brewing your coffee. Note that most competitors roast in advance since that is the way they typically have sampled their roasted coffee. It is interesting, however, to explore how the coffee evolves with storage time after roasting. In Ethiopia, the birthplace of coffee, coffee is roasted and immediately brewed, yielding a fantastic cup (albeit typically mixed with many spices). According to the SCA, coffee should be allowed to rest for at least 8 hours before cupping, but feel free to explore!

6) **Cold Water:** You are only allowed to use **regular cold tap water**, not hot tap water, as your starting point for your brew. The kettles should be empty and cool to the touch before you start. An interesting aspect is to allow competitors to use their own "designed" water for brewing, as discussed in Lab 7 on Water Quality.

7) **No Flames:** Absolutely no open flames are allowed in the lab, so no stoves or burners.

8) **Negative Infinity:** Clever students will note that if their process uses *zero* electrical energy, then their final score will be infinity (because any tasting score divided by zero is infinity). One example of a zero electrical energy process would be to pour cold water over whole green beans. Although this is a clever idea in terms of energy minimization, note that *negative* tasting scores are possible (because of the 'balance' category, which allows scores between –10 and +5. So, in this competition it is likely that such a low energy process would actually yield a score of *negative infinity* and therefore *last place* in the competition (assuming you were the only group silly enough to try this).

9) **Creativity:** Most importantly, remember that good engineers think of new ways of doing things… we encourage creative designs!

Final Design Lab Report

Your final lab report will serve as a summary of your engineering design for the contest. It must include the following material.

1) A process flow diagram that shows your unique process. There must be enough detail that somebody else can replicate your process. Remember, even if you didn't directly measure a material stream, you can calculate it using an appropriate mass balance.
2) Your overall mass balances for water and solid (coffee). These can be shown on the process flow diagram. Make sure you include all waste streams!
3) Your type of roast(s). What beans did you use, and how did you roast them?
4) Your energy usages. Clearly describe the logic you followed to attempt to minimize energy costs.
5) Your TDS and PE for your one of your final design brews, or your competition brew from the contest. How close were you to the ostensible ideal range?
6) An overview of the sensory evaluations of your final brew as judged your group and by others in the blind tasting.
7) A summary of anything you would do differently in a (hypothetical) future design contest to improve your coffee taste or minimize your energy usage.

Pictures taken while you were roasting and brewing in lab are recommended but not strictly necessary.

Don't worry about having the perfect design yet! The next few labs are intended for you to hone your roast and brew.

Lab 8 – First Design Trials: Optimizing Strength & Extraction

Objectives: The main goal of this lab is for you and your group to begin designing your process, with an emphasis on optimizing the strength and extraction of your brew.

Equipment:

☐ Brewer(s) ☐ Roaster(s) ☐ Digital refractometer ☐ Kill-a-Watt meter

Activities:

☐ At least three brews, completely quantified (TDS, PE, energy)

☐ At least two roasts, with energy measurements

Report:

☐ Data for three brews

☐ A completed energy scoresheet

☐ Comparison of predicted vs. calculated PE for one brew

☐ Paragraphs discussing tasting notes and proposed gameplan for next design trials

Background

If you think back on the previous ten labs, in each experiment we tried varying some experimental condition and then seeing what happened, with some insight provided by fundamental principles (e.g., conservation of mass, flux, or pressure). In other words, we were performing *analysis* of how the brewing method impacted coffee quality. Now we need to change our focus to *design*, where we use the knowledge gained via analysis to create a process that satisfies design goals.

In our case, we want to make the best-tasting coffee using the least amount of energy. As discussed back in Lab 6, a great deal of sensory analysis experimentation has shown that people prefer the taste of coffee in a pretty narrow range of strength (TDS from about 1 to 1.5 %) and extraction (PE from about 14 to 26%, although different people prefer different ranges of PE and TDS). Refer again to the coffee brewing control chart (next page) that shows the different sensory evaluations one gets from brews prepared to different strengths and extractions. If your coffee tastes too bitter, it is probably over-extracted (PE > 22%), but if it tastes sour or vegetal it is probably under-extracted (PE < 18%). The TDS of the brew tends to amplify the intensity of the extracted flavors for most sensory attributes, but counterintuitively some prized attributes increase in intensity at *lower* values of TDS: sweetness and tea-like/floral notes are maximized at low TDS and low or high PE, respectively. Refer to the articles described in the Further Reading section (Frost et al. 2020 and Batali et al. 2020) for more details on the underlying science.

Heretofore we made some coffee and then measured the TDS and PE after the fact. But what do we do if we want to predict these values *before* we do the experiment, i.e., if we want to design the process to yield coffee in the ideal range of TDS and PE? To tackle this problem, we will combine several of our analytical results into a form that will be convenient for predicting the outcome of a brew.

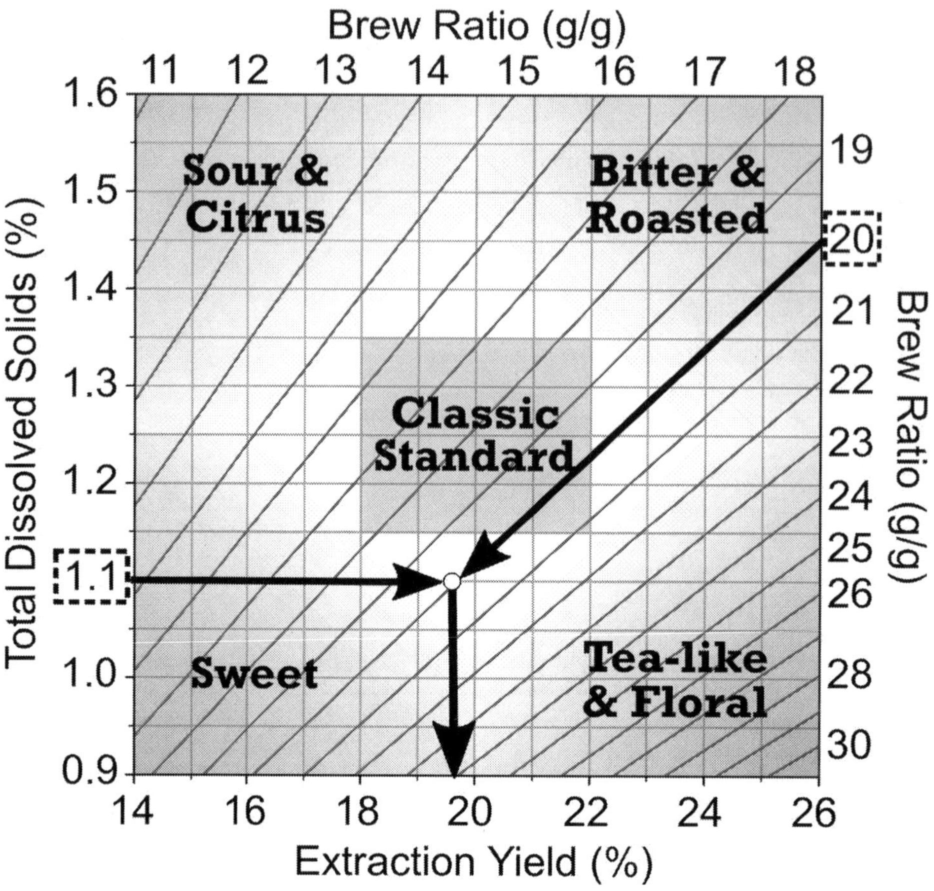

Recall that our mass balance on the coffee solids yielded an expression for the PE,

$$PE = TDS \times \frac{m_{brew}}{m_{dry\ grounds}}. \tag{8.1}$$

Also recall that our mass balance on the water yielded a prediction for the actual mass of the brew,

$$m_{brew} = m_{feed} - \left(R_{abs} \times m_{dry\ grounds}\right). \tag{8.2}$$

Substitution of our expression for m_{brew} into our expression for PE yields

$$PE = TDS \times (R_{brew} - R_{abs}), \tag{8.3}$$

where R_{brew} is the brew ratio defined as the mass of hot water fed in over the mass of dry grounds, i.e., $R_{brew} = m_{feed}/m_{dry\ grounds}$. There is a hugely important design implication of equation (8.3): if you want to obtain a certain value of PE, you *cannot* independently choose a brew ratio and a TDS value. Look at the highlighted point in the control chart above. Here a brew ratio of 20 was used, and the brew had a measured TDS of 1.10%. At the intersection of the corresponding horizontal line (for TDS) and diagonal line (for brew ratio), there is only one possible corresponding PE of 19.6%, as indicated by the vertical downward line.

In other words, once you choose a brew ratio, there is only one possible TDS value that is consistent with the desired PE. This idea is captured in the diagonal lines superimposed on the coffee brewing control chart, which show the possible values of TDS and PE for the specified brew ratio. As you can see, there are only a narrow range of brew ratios that can yield brews within the classic standard zone.

Once you choose a brew ratio, though, how do you specify the TDS? Again, the insights you gained via engineering analysis in the previous labs will now help you design the process. If your TDS (and hence PE) are too small and your brew is under-developed, then you likely had insufficient mass transfer during your extraction. This could be due to a grind that was too coarse, or an extraction time that was too short, or a water temperature that was too low. Likewise, if your TDS and PE are too large and your brew is bitter, you had too much transfer during your extraction: too fine a grind, too long an extraction time, or too high a water temperature. During your design trials today, don't try things randomly: use the insight you gained on mass transfer, as well as equation (8.3) above, to improve your brewed coffee.

An important question is: should you strive to get your brew within the "classic standard" zone of the coffee brewing control chart? The short answer is: probably not. Although for years coffee industry professionals were taught that it is crucial to get their brew inside that zone, recent consumer preference testing has shown significant variations in what people actually like to drink. In particular, Cotter et al. (2021) showed that black coffee drinkers could be segmented into two cohorts: one cohort strongly preferred coffees brewed to the bottom left corner of the chart (labeled as "sweet"), while the other cohort equally preferred either the top left corner (labeled "sour & citrus") or the far-right side of the graph (between "bitter & roasted" and "tea-like & floral").

Of course, don't forget that the quality of the brew depends on the quality of the chemical reactions that occurred during roasting. If you under- or over-roasted your beans, it won't matter how well you do the extraction, because you'll just be extracting out unpleasant tasting coffee. **You could extract very, very burnt roasted beans (or even black charcoal) to the 'classic standard' range of 1.3% TDS and 20% PE... but it won't taste very good!** Remember as well, coffee beans are biological products that can vary from lot to lot. What worked for one type of beans won't necessarily work with a different type of beans. For example, the green beans might have different concentrations of sucrose depending on how they were grown, or perhaps they've been stored for different lengths of time and have degraded differently. Coffee is a moving target! You need to adjust your aim, i.e., refine your design, based on your experimental observations for any particular batch of beans. Your ultimate evaluation of brew quality should be based on your sensory evaluation of the taste.

Summary of 1st Design Trial Activities

Unlike the previous labs, we intentionally leave the design trial activities open ended. Use the time to work with your group on making and testing your own process for roasting and brewing the best tasting coffee. You are welcome to use any combination of equipment available in the lab, and you can use the time as you think best. We urge you to think carefully, however, about the following questions.

1) What are the TDS and PE of your brews? Are they close to the classic standard range? Do your own sensory evaluations agree with the taste descriptions on the coffee brewing control chart?

2) What is the best roasting procedure? Make sure you perform at least a couple roasts toward your ideal roast.

3) How much energy is each step using? Is your design energy efficient, or an energy hog? Make sure you get the required data to fill out the sample "energy scoring sheet," which is similar to what will be used in the design competition. (If you're using an automatic drip brewer like Mr. Coffee, don't worry about collecting temperature data since you have no control over it.)

Data for 1st Design Trials

1st Roast Coffee type: ____

Initial roaster settings: ____________ Changes to roaster settings: ____________

Mass of green beans: _______ *grams* Mass of roasted beans: _______ *grams*

First crack: _____ *mins* Second crack: _____ *mins* Roast time: _____ *mins*

Color: ______________ Energy usage: _______ *kW-hr*

Notes: ____

Brewing method: ______________ Filtration method: ______________

Mass of hot water: _______ *grams* Mass of grounds: _______ *grams* Ratio: _______

Water Temp: ______ °C Energy usage: _______ *kW-hr*

Extraction time: _____ *minutes* Grind size: __________

Mass of brew: _______ – _______ = _______ *grams* (filled cup – empty cup)

TDS: _______ *%* PE: _______ × _______ ÷ _______ = _______ *%*

Other data:

____

____

____

____

Sensory Evaluations:

____

____

____

____

Data for 1st Design Trials

2nd Roast Coffee type: ____

Initial roaster settings: ____________ Changes to roaster settings: ____________

Mass of green beans: _______ *grams* Mass of roasted beans: _______ *grams*

First crack: _____ *mins* Second crack: _____ *mins* Roast time: _____ *mins*

Color: ______________ Energy usage: _______ *kW-hr*

Notes: ____

Brewing method: ______________ Filtration method: ______________

Mass of hot water: _______ *grams* Mass of grounds: _______ *grams* Ratio: _______

Water Temp: ______ °C Energy usage: _______ *kW-hr*

Extraction time: _____ *minutes* Grind size: __________

Mass of brew: _______ – _______ = _______ *grams* (filled cup – empty cup)

TDS: _______ *%* PE: _______ × _______ ÷ _______ = _______ *%*

Other data:

____

____

____

____

Sensory Evaluations:

____

____

____

Data for 1st Design Trials

3rd Roast Coffee type: ______________________________

Initial roaster settings: ____________ Changes to roaster settings: ____________

Mass of green beans: ________ *grams* Mass of roasted beans: ________ *grams*

First crack: ______ *mins* Second crack: ______ *mins* Roast time: ______ *mins*

Color: ______________ Energy usage: ________ *kW-hr*

Notes: ____

Brewing method: ______________ Filtration method: ______________

Mass of hot water: ________ *grams* Mass of grounds: ________ *grams* Ratio: ________

Water Temp: _______ °C Energy usage: ________ *kW-hr*

Extraction time: ______ *minutes* Grind size: __________

Mass of brew: ________ – ________ = ________ *grams* (filled cup – empty cup)

TDS: ________ *%* PE: ________ × ________ ÷ ________ = ________ *%*

Other data:

____

____

____

____

Sensory Evaluations:

____

____

____

____

Sample Energy Scoring Sheet

TEAM NAME: ______________________________

If your brew uses beans from just one roast, then fill out only "Roast A" and put zero for "Roast B." If you use three or more roasts, append the data and calculate the total energy of roasting appropriately.

Roast A (bean type): _______________

Mass of green beans: ________ *grams*

Mass of roasted beans: ________ *grams*

Total energy of roast: ________ *kW-hr*

Energy to roast per gram roasted bean:

______ ÷ ______ = _______ *kW-hr/gram*

Total mass of beans used: _______ *grams*

Energy per gram × total grams used:

______ × ______ = _______ *kW-hr*

Roast B (bean type): _______________

Mass of green beans: ________ *grams*

Mass of roasted beans: ________ *grams*

Total energy of roast: ________ *kW-hr*

Energy to roast per gram roasted bean:

______ ÷ ______ = _______ *kW-hr/gram*

Total mass of beans used: _______ *grams*

Energy per gram × total grams used:

______ × ______ = _______ *kW-hr*

Total energy for roasted coffee actually used in your brew (Roast A + Roast B):

________+ ________ = ________ *kW-hr*

Brewing method(s): ___________________________________

Total mass of water heated: ________ *grams* *(make sure the kettle is cool and empty at start)*

Initial Water Temp: ________ °C Final Water Temp: ________ °C

Energy used to heat water: ________ *kW-hr*

Mass of brew: ________ – ________ = ________ *grams* *(filled carafe– empty carafe)*

TDS: ________ *%* PE: ________ × ________ ÷ ________ = ________ *%*

RANDOM NUMBER GENERATOR: TARGET BREW MASS = ______ grams

Deficient mass: _______ – _______ = _______ *grams*

Energy penalty: N/A *grams* × 0.005 *kW-hr/gram* = N/A *kW-hr*

(Note: don't worry about the energy penalty in this lab since you're not making a whole liter.)

Total energy used to produce your coffee (roast energy + water energy + penalty):

________+________+ 0 = ________ *kW-hr*

More Roasting Data

Coffee type: ______________________________

Initial roaster settings: ____________ Changes to roaster settings: ____________

Mass of green beans: ________ *grams* Mass of roasted beans: ________ *grams*

First crack: ______ *mins* Second crack: ______ *mins* Roast time: ______ *mins*

Color: ______________ Energy usage: ________ *kW-hr*

Notes: ____

Coffee type: ______________________________

Initial roaster settings: ____________ Changes to roaster settings: ____________

Mass of green beans: ________ *grams* Mass of roasted beans: ________ *grams*

First crack: ______ *mins* Second crack: ______ *mins* Roast time: ______ *mins*

Color: ______________ Energy usage: ________ *kW-hr*

Notes: ____

Coffee type: ______________________________

Initial roaster settings: ____________ Changes to roaster settings: ____________

Mass of green beans: ________ *grams* Mass of roasted beans: ________ *grams*

First crack: ______ *mins* Second crack: ______ *mins* Roast time: ______ *mins*

Color: ______________ Energy usage: ________ *kW-hr*

Notes: ____

Lab Report

By your specified due date, each group will submit a lab report that includes:

(1) For at least three brews, state (i) the brewing method, (ii) grind level, (iii) TDS and PE (show the calculations), (iv) location on the control chart (take a picture of the chart and mark where your three brews were), and (v) tasting notes.

(2) A completed "energy scoring sheet" for whichever process you think is most promising for the competition. (A photo or scan of an energy score sheet is fine.)

(3) For at least one brew, a comparison of your calculated PE using Equation (8.1) versus the PE predicted by equation (8.3). How similar or dissimilar are the values?

(4) Finally, a paragraph that describes your main findings today, and a brief game plan for what you will test in your design next week. What logic or experimental data informed your design choices?

Lab 8 Bonus Box – How is Coffee Decaffeinated?

The Lab 2 Bonus Box was titled "Caffeine the Wonder Drug," but what if you don't want caffeine in your coffee? Some folks are very sensitive to caffeine. Others might like to have an after-dinner cup of coffee deliciousness, but they find it difficult to sleep afterwards. Everyone has heard of decaffeinated coffee, but how the caffeine removed?

As we mentioned earlier, caffeine is a natural molecule made by *Coffea* plants as a defense mechanism against insects. A lot of folks think that dark roasts have more caffeine than light roasts, but the amount of caffeine is really dictated by the bean – not the roast level. Caffeine is an alkaloid (a nitrogen containing compound) and it is pretty impervious to the roasting process, so a dark or light roast of the same beans has about the same caffeine content. The differences in the amount of caffeine in a typical brew are mainly due to the amount of caffeine that happened to be in the bean to begin with, and how efficiently you extracted it during the brewing process.

Green beans are always decaffeinated **before** roasting. The first step invariably involves steam. The solubility of caffeine in water is highly sensitive to temperature: at room temperature the solubility is about 2 grams per 100 grams water, but it increases dramatically to 66 grams per 100 grams boiling water. As a result, the first step in any decaffeination process is to "swell" the green coffee beans with steam. The beans actually swell and increase in size about 50% during steaming. This opens up the pores in the beans, making the caffeine more accessible. The steaming also helps to solvate and mobilize the caffeine molecules.

After the steaming step, there are four main methods for decaffeinating coffee, which are separated into "solvent-based" processes using chemicals and "non-solvent based" processes using water. (Actually, strictly speaking all of them use solvents since technically water is a solvent as well… but even though water is a chemical, using the phrase "chemical free" is a matter of semantics that is important to many consumers.) In the direct solvent process, a liquid that has a high caffeine solubility but a lower "coffee stuff" solubility (everything else that isn't caffeine) is contacted with the steamed green coffee beans so that the caffeine is selectively dissolved (or "solubilized") into the solvent. The solvents used in modern processes are typically ethyl acetate or methylene chloride. If ethyl acetate is used, the process is often termed "natural decaffeination" because ethyl acetate is found in fruit and can also be obtained from fermentation of sugar cane; it is considered more "natural" than chemically synthesized methylene chloride (called dichloromethane in Europe). Despite the natural name, the ethyl acetate used in decaffeination is industrially produced because it is much cheaper to synthesize with chemical reactions than trying to extract the chemical from fruit or produce it by fermenting sugar. Regardless, ethyl acetate and methylene chloride are both mildly toxic so you don't want them in your cup of coffee. Fortunately, only a few ppm of either solvent remains at the end of the decaffeination process, and essentially none remains after the beans are roasted. (Recall the discussion on ppm in Bonus Box 6.)

To extract the caffeine, the steamed coffee beans are simply soaked in the solvent for about 10 hours, the solvent is removed, and the green decaffeinated beans are re-steamed to remove any residual solvent. This process is called a "direct solvent" method because the green beans directly contact the solvent. In the "indirect solvent" method, the green beans are steamed and soaked in water, and the water is then separated from the beans. The caffeine-rich water is then mixed with the solvent to extract the caffeine from the water into the solvent. This procedure keeps the beans from touching the solvent directly but also tends to cause more loss of flavor compounds.

To help address this problem, and to avoid use of chemical solvents, alternative approaches that "recycle" the water have been developed. One such "non-solvent" based approach is the "Swiss water process." Named because it was developed in Switzerland (not because it uses Swiss water). In this process the green beans are soaked in near-boiling water to remove the caffeine, which also inadvertently removes other soluble flavor molecules. The water is collected and the caffeine is removed by filtering through "activated charcoal." The charcoal has a porosity (average hole size) that allows the flavor molecules to get through but traps the larger caffeine molecules. After removing the caffeine, the green coffee "flavored" water is reused to remove the caffeine from the next batch of green beans. The first batch of beans are "sacrificial beans" that help saturate the water with those important water soluble flavor molecules that come out in the hot water along with the caffeine. After the caffeine is removed by the charcoal, the second batch of beans "sees" a liquid that already has lots of flavor molecules, so there is a smaller concentration difference to serve as a driving force for the molecules to leave the beans – except for the caffeine, which will rapidly solubilize into the caffeine-free water. Reusing the water in this fashion helps prevent the extraction from the next batch of beans to be processed and so on.

The fourth and final decaffeination technique involves something called "super critical carbon dioxide," which is some pretty interesting stuff. Carbon dioxide (CO_2) is a product of respiration – we exhale this gas as a by-product of using energy – and it is a product from reactions that burn or release energy (including coffee roasting and combustion engines like in cars). Caffeine and most other liquid and solid chemicals are not very soluble in CO_2 gas. However, if you compress CO_2 gas to a high enough pressure (around 250 atmospheres), it can remain in a sort of gas phase, but almost have the density of a liquid, which is why it has the moniker "super critical." Even though caffeine isn't all that soluble in super critical CO_2, it is much more soluble that the other flavor molecules. After steaming the green beans, the wet, swelled-up beans are contacted with the super critical CO_2 in a "counter current" fashion: beans enter the top of the column and super critical CO_2 flows up from the bottom. As you learned in Lab 6, mass transfer depends on a concentration gradient. By having "fresh" CO_2 contact the already mostly caffeine extracted green beans that have been in contact with CO_2 since the top of the column, the concentration gradient can be enhanced over the length of the column to improve the caffeine transfer. The CO_2 is collected from the top of the tower and contacted with a water spray to leach the caffeine from the CO_2. The CO_2 is then recycled, while the nearly pure caffeine can be sold.

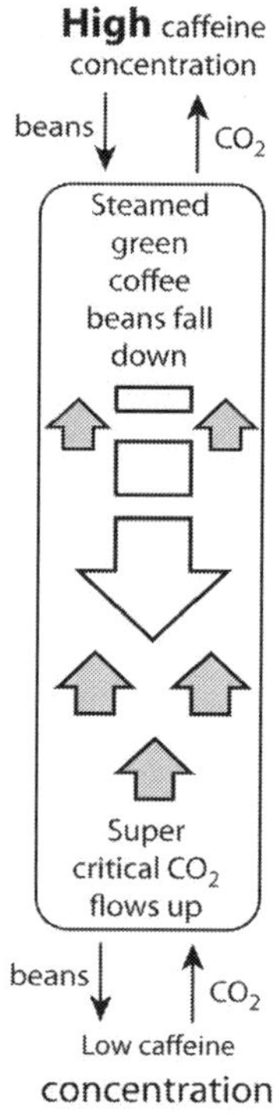

Having said all this, we should note that many folks (incorrectly) believe that there simply is no such thing as a good cup of decaffeinated coffee. The thinking is that no matter what decaffeinating process is used, there is invariably some removal of molecules, which will affect flavor. Whether the loss is good or bad, however, depends on the molecules – sometimes decaffeination can remove molecules that cause flavor defects! Overall, the situation for decaf coffee is improving. In the past much decaffeinated coffee was not that good to begin with – in other words, garbage in, garbage out. With modern techniques, and if you start with a high-quality coffee and use careful processing, a cup of delicious high-quality decaf can be obtained. Decaffeinated green beans are readily purchased online – get some and try your own experiments!

Lab 9 – Second Design Trials: Scaling Up to One Liter of Coffee

Objectives: The main goal of this lab is to plan how to "scale up" your process efficiently to a larger volume, and to continue optimizing your roast and brew for the competition!

Equipment: -

☐ Brewer(s) ☐ Roaster(s) ☐ Digital refractometer ☐ Kill-a-Watt meter

Activities:

☐ At least three brews, completely quantified (TDS, PE, energy)

☐ Prepare more roast(s) for the final design competition, with energy measurements

☐ Consideration of time and energy usage

☐ Pictures and/or video as you desire for the final design video

Report:

☐ A completed energy scoresheet

☐ Estimates of how much water, coffee, energy, and time are needed for 1 liter

☐ Paragraphs discussing your proposed design and logic for design choices

Background

There are many processes that work beautifully at "small" scales but are challenging (or impossible) to implement at "large" scales. For example, a chemist might devise an elegant chemical reaction process for synthesizing some desired compound. Even though the chemist can get the reaction to work in the lab, there might be many engineering challenges to overcome before the reaction process can be used industrially. If the reaction involves too many steps, or requires too many expensive catalysts, or only works at impractically high temperatures, or yields too many waste products, then we might conclude that the reaction can't be "scaled up" efficiently.

The same logic applies to coffee. Perhaps you have perfected a process for making a small cup of coffee that tastes great… but that doesn't necessarily mean you can use the same process to make a larger volume of coffee. Perhaps the process requires an unacceptably large amount of energy; perhaps the taste of the coffee changes when you try to brew over a larger volume; perhaps the process is so slow that it simply requires too much time.

The main goals of this design trial are (1) to finalize the roast profile for the beans you will be using in the final competition and (2) to plan how to "scale up" your brewing process efficiently to a larger volume. It is crucial that you plan how much water, how much ground coffee, how much energy, and how much time are required. Don't waste time making a whole liter today – instead, optimize your design and plan how it will work during the competition!

It is crucial – absolutely crucial – that when you and your group walk into the design contest, you know how many coffee grounds you will need, how much hot water you will need, and how much energy it should all require. The materials already discussed in previous labs will help us be quantitative. First, let's think about how much water we'll need. We can calculate exactly how much water we'll need, by rearranging equation (3.5) in Lab 3, yielding

$$m_{water} = m_{brew} + (R_{abs} \times m_{grounds}), \tag{9.1}$$

where m_{water} is the total amount of water you need to feed into our brewer. (Make sure you calculate the quantitative value of R_{abs} for your preferred brewing technique!) How much energy will it cost to heat up this much water? Well, think back to Lab 5, where we showed how the energy is proportional to the total mass, the heat capacity, and the desired temperature increase. Substitution of equation (9.1) above into equation (5.3) from Lab 5 yields

$$E_{brew} = [m_{brew} + (R_{abs} \times m_{grounds})] \times C_p \times \Delta T. \tag{9.2}$$

Think carefully about these equations, and how your choices for $m_{grounds}$ and m_{water} affect your possible ranges for TDS and PE (and the corresponding sensory qualities).

Timing is also important. Obviously, you don't want to run out of time while brewing... you'll suffer a huge energy penalty if you don't brew enough coffee in time or even be disqualified. Also recall your experience in Lab 4, where the pH of the coffee changed with time, and the discussion in Lab 7 of how brewing flow rate allows the coffee grounds to continue to extract, so the flavor changes with time. Your brew will probably degrade in quality if you finish the brews too quickly and let them sit around for half an hour before anybody tastes them. You'll be combining your brews in a 1 L insulated thermal carafe during the competition, so pH likely won't be too affected, but extraction of any coffee grounds that you let enter the carafe will continue. So, a key goal for this lab is to time how long each step in your process takes, and to estimate the total time required to make a liter of brew. As before, you also need to complete your energy scoresheet.

Summary of 2nd Design Trial Activities

As with the first design trials, we intentionally leave the lab activities open-ended, so that you can work with your group on your own process design. Try at least a couple different overall designs, preferably more. This lab is also a great time to obtain some video for the video project. Think carefully about the following questions.

1) If you want to end up with precisely m_{brew} grams of brewed coffee, how much water will you need to heat? You don't want to waste energy heating up water that you don't use. You also don't want to end up short of this amount. Think back to Lab 3... what information do you need to plan a process yielding precisely a desired amount of coffee? Will the data you obtained in Lab 3 apply to your process? (Hint: what is R_{abs} for your brewing method?) If not, what experiment and measurements could you do?
2) Recall that you'll have 45 minutes to brew about 1 liter of coffee. If your process involves any repeated slow steps, will you have sufficient time to make a whole liter? Contrariwise, if your process is really fast, will your brew be sitting around for a long time before anybody tastes it? Record the total amount of time it takes you to complete a brew. Note that extra brewing equipment will be available for the competition.
3) Keep refining your roasts... you are getting very close to the competition!
4) Make sure you record the energies used in every step. You will need to fill out an energy scoresheet like the one that you will use in the actual competition.
5) Also get the TDS and PE but remember those are only guides... your perceived taste evaluation is the most important!

Data for 2nd Design Trials

1st Roast Coffee type: ____

Initial roaster settings: ____________ Changes to roaster settings: ____________

Mass of green beans: _______ *grams* Mass of roasted beans: _______ *grams*

First crack: _____ *mins* Second crack: _____ *mins* Roast time: _____ *mins*

Color: ______________ Energy usage: _______ *kW-hr*

Notes: ____

Clock time when started prepping to brew: __________

Brewing method: ______________ Filtration method: ______________

Mass of hot water: _______ *grams* Mass of grounds: _______ *grams* Ratio: _______

Water Temp: ______ °C Energy usage: _______ *kW-hr*

Extraction time: _____ *minutes* Grind size: __________

Mass of brew: _______ – _______ = _______ *grams* (filled cup – empty cup)

Clock time when brew finished: __________ Time required: _______ *minutes*

TDS: _______ % PE: _______ × _______ ÷ _______ = _______ %

Other data:

____

____

____

Sensory Evaluations:

____

____

____

____

Data for 2nd Design Trials

2nd Roast Coffee type: ______________________________

Initial roaster settings: ______________ Changes to roaster settings: ______________

Mass of green beans: ________ *grams* Mass of roasted beans: ________ *grams*

First crack: ______ *mins* Second crack: ______ *mins* Roast time: ______ *mins*

Color: ________________ Energy usage: ________ *kW-hr*

Notes:____

Clock time when started prepping to brew: ___________

Brewing method: ________________ Filtration method: ________________

Mass of hot water:________ *grams* Mass of grounds:________ *grams* Ratio:________

Water Temp: _______ °C Energy usage: ________ *kW-hr*

Extraction time: ______ *minutes* Grind size: ___________

Mass of brew: ________ – ________ = ________ *grams* (filled cup – empty cup)

Clock time when brew finished: ___________ Time required: ________ *minutes*

TDS: ________ *%* PE: ________ × ________ ÷ ________ = ________ *%*

Other data:

____

____

____

Sensory Evaluations:

____

____

____

____

Data for 2nd Design Trials

3rd Roast Coffee type: ____

Initial roaster settings: ____________ Changes to roaster settings: ____________

Mass of green beans: _______ *grams* Mass of roasted beans: _______ *grams*

First crack: _____ *mins* Second crack: _____ *mins* Roast time: _____ *mins*

Color: ______________ Energy usage: _______ *kW-hr*

Notes: ____

Clock time when started prepping to brew: __________

Brewing method: ______________ Filtration method: ______________

Mass of hot water: _______ *grams* Mass of grounds: _______ *grams* Ratio: _______

Water Temp: ______ °C Energy usage: _______ *kW-hr*

Extraction time: _____ *minutes* Grind size: __________

Mass of brew: _______ – _______ = _______ *grams* (filled cup – empty cup)

Clock time when brew finished: __________ Time required: _______ *minutes*

TDS: _______ % PE: _______ × _______ ÷ _______ = _______ %

Other data:

____

____

____

Sensory Evaluations:

____

____

____

____

Sample Energy Scoring Sheet

TEAM NAME: ______________________________

If your brew uses beans from just one roast, then fill out only "Roast A" and put zero for "Roast B." If you use three or more roasts, append the data and calculate the total energy of roasting appropriately.

Roast A (bean type): _______________

Mass of green beans: ________ *grams*

Mass of roasted beans: ________ *grams*

Total energy of roast: ________ *kW-hr*

Energy to roast per gram roasted bean:

______ ÷ ______ = _______ *kW-hr/gram*

Total mass of beans used: _______ *grams*

Energy per gram × total grams used:

______ × ______ = _______ *kW-hr*

Roast B (bean type): _______________

Mass of green beans: ________ *grams*

Mass of roasted beans: ________ *grams*

Total energy of roast: ________ *kW-hr*

Energy to roast per gram roasted bean:

______ ÷ ______ = _______ *kW-hr/gram*

Total mass of beans used: _______ *grams*

Energy per gram × total grams used:

______ × ______ = _______ *kW-hr*

Total energy for roasted coffee actually used in your brew (Roast A + Roast B):

________+ ________ = ________ *kW-hr*

Brewing method(s): ____________________________________

Total mass of water heated: ________ *grams* *(make sure the kettle is cool and empty at start)*

Initial Water Temp: ________ °C Final Water Temp: ________ °C

Energy used to heat water: ________ *kW-hr*

Mass of brew: ________ – ________ = ________ *grams* *(filled carafe– empty carafe)*

TDS: ________ % PE: ________ × ________ ÷ ________ = ________ %

RANDOM NUMBER GENERATOR: TARGET BREW MASS = _____ grams

Deficient mass: 925 – _______ = _______ *grams*

Energy penalty: N/A *grams* × 0.005 *kW-hr/gram* = N/A *kW-hr*

(Note: don't worry about the energy penalty in this lab since you're not making a whole liter.)

Total energy used to produce your coffee (roast energy + water energy + penalty):

________+________+ 0 = ________ *kW-hr*

More Roasting Data

Coffee type: ______________________________

Initial roaster settings: ____________ Changes to roaster settings: ____________

Mass of green beans: _______ *grams* Mass of roasted beans: _______ *grams*

First crack: _____ *mins* Second crack: _____ *mins* Roast time: _____ *mins*

Color: ______________ Energy usage: _______ *kW-hr*

Notes: ____

Coffee type: ______________________________

Initial roaster settings: ____________ Changes to roaster settings: ____________

Mass of green beans: _______ *grams* Mass of roasted beans: _______ *grams*

First crack: _____ *mins* Second crack: _____ *mins* Roast time: _____ *mins*

Color: ______________ Energy usage: _______ *kW-hr*

Notes: ____

Coffee type: ______________________________

Initial roaster settings: ____________ Changes to roaster settings: ____________

Mass of green beans: _______ *grams* Mass of roasted beans: _______ *grams*

First crack: _____ *mins* Second crack: _____ *mins* Roast time: _____ *mins*

Color: ______________ Energy usage: _______ *kW-hr*

Notes: ____

Lab Report By the specified due date, each group will submit a lab report that includes:

(1) A completed "energy scoring sheet" for whichever process you think is most promising for the competition. (A photo or scan of an energy score sheet is fine.)

(2) An estimate of the total amount of water and ground coffee that you will need to make 1 liter of brewed coffee at the competition, along with a corresponding estimate of how much energy will be used. Show all your calculations (with units).

(3) An estimate of how much time will be required to brew a complete liter, based on your measurements of time required here for individual brews.

(4) A paragraph that describes the brewing method, roast information, approximate grind level, your TDS and PE, where you are on the ideal brewing chart, your tasting notes, and any other information you feel is relevant.

(5) Another paragraph that describes *why* you have selected your potential process for the competition. What logic or data informed your design choices?

Lab 9 Bonus Box – How is instant coffee made?

In the United States we don't really think that much about instant coffee. Only a piddling 3% of the coffee consumed in the US is instant, a shockingly small amount given that Americans are well known to be willing to pay a premium for convenience and speed. In contrast, more than 1/3 of the coffee consumed worldwide is instant coffee. Moreover, the consumption of instant coffee is increasing rapidly – it has tripled over the past 15 years. The percentage of instant coffee sold in the US per capita is close to the smallest in the world.

It's not that Americans don't drink coffee. The average American drinks 2.6 cups of coffee a day. Some of the discrepancy is due to the surge in coffee pods in the US. Those single servings of coffee that can be brewed with the push of a button now make up 25% of the American market. They are fast and convenient, but not "instant" coffee.

Elsewhere in the world, instant coffee has a huge footprint. Nestlé (also known as Nescafé) buys about 15% of all the coffee produced each year. Although "instant" coffee was first reportedly developed around 1771 in Britain, it wasn't till 1938 that Nestlé developed a reasonably palatable version. During that time, Brazil was producing a large excess of coffee, and the Brazilian government started an initiative to figure out what to do with the excess. Nestlé came up with a method of drying coffee with carbohydrates that reportedly was drinkable. The timing was perfect. Shortly thereafter, the advent of World War II solidified the place of instant coffee as a quick and convenient method to fuel soldiers.

Today, instant coffee is mostly made by two processes: spray drying or freeze drying. In spray drying, concentrated brewed coffee is sprayed to make a fine mist and dried with hot air. To concentrate the coffee, it is brewed at a high coffee to water ratio. Some of the water is then evaporated off to yield a 50% or greater strength (i.e., TDS). That's pretty concentrated as a typical cup of coffee has a concentration of about 1.3%. The concentrated mist is then sprayed at the top of a tower, and hot air is blown upwards from the bottom of the column (similar to the counter-current flow for caffeine extraction in the Lab 11 Bonus Box). At the bottom of the column the dried coffee particles are collected. Those particles are agglomerated (stuck together with a little water) to make larger particles. Sometimes carbohydrates are added to increase the volume and make it look more like ground coffee, and sometimes fresh, finely ground coffee is added for more flavor. Agglomeration is how they make instant coffee powder look more like ground coffee when you open the package. Once the particles have the size and look that the manufacturer wants, it's packaged and ready to go.

In freeze drying, the concentrated coffee is first frozen before drying. It sounds a little like iced coffee, but the starting point is coffee ice cubes. These cubes are broken into small pieces and then put into a vacuum chamber and dried. In the Lab 5 Bonus Box, the phase diagram of water is shown. As you may recall, at pressures below 0.006 atmospheres when you heat ice it doesn't melt but instead vaporizes – the water molecules go directly from the solid phase to the gas phase. This process is called sublimation or "freeze drying." Freeze drying costs quite a bit more than spray drying, but most high-quality instant coffee manufacturers use freeze drying because this process retains more of the aroma and flavor molecules than spray drying, resulting in a better tasting instant brew. In fact, those volatile aroma and flavor molecules are so critical to the quality that there are even processes that collect the gases during the extraction (brewing) phase and evaporation phase to try to add them back into the instant coffee. A lot of work to make an instant drink!

Lab 10 – Design Competition & Blind Taste Panel

Objective: To obtain fame, glory, and bonus points by winning the grand prize!

Equipment:

☐ Brewer(s) ☐ Kill-a-Watt meter ☐ Large insulated carafes ☐ Gong (optional!)

Activities:

☐ Part A – Brew the required mass of coffee and fill out energy scoresheet in 45 minutes

☐ Part B – Measure TDS of your final brew, and submit your carafe and scoresheet

☐ Part C – Blind taste test of everybody's coffee

Report:

☐ Video report of your final design

Summary of Contest Activities

There are two parts to the design contest. First, you'll have 45 minutes to brew almost one liter of coffee (up to 925 grams to be exact) and fill out your energy scoresheet. You will also measure your final TDS. Then, everybody will participate in the blind tasting. Before you leave, you will know the winning group!

Part A – Brewing 1 Liter of Coffee

As soon as you arrive in lab, wash your hands as usual, and feel free to organize your station and get your roasted beans out. **Empty and rinse your kettle with cool water if it's not already cool to the touch.** DO NOT, however, begin to weigh anything, or grind anything, or heat any water. The contest coordinator will first give an overview of the contest guidelines and rules and then ask for a volunteer to perform the ceremonial ringing of the gong. As soon as the gong rings, your group has 45 minutes to make your coffee! Each group will be provided with a thermally insulated carafe that holds up to 1 liter of coffee, but to avoid spills or overflows you will only brew 925 grams.

Before the gong rings, the contest coordinator uses a random number generator to specify a target mass between 600 and 925 grams. This requires the competitors to rapidly scale their process design in real time for the competition.

Once the gong rings, zero your Kill-a-Watt meter. Any device you use **must** be plugged into the meter and not re-zeroed until you complete your brew. While you are brewing, continue to measure your energy usage, and completely fill out your energy scoring sheet. If you have extra beans, it's a good idea to do a test grind to make sure the grind size is what you want for the competition, and to flush out any residual grounds. It's probably also a good idea to rinse out your carafe thoroughly before you put your brew in. If you rinse it with warm water (why might you want to do that?) make sure you include the energy usage!

Design Competition Energy Scoring Sheet

TEAM NAME: ____

Station number: ____________ **Section number:** ___________

If your brew uses beans from just one roast, then fill out only "Roast A" and put zero for "Roast B." If you use three or more roasts, append the data and calculate the total energy of roasting appropriately.

Roast A (bean type): ________________

Mass of green beans: ________ *grams*

Mass of roasted beans: ________ *grams*

Total energy of roast: ________ *kW-hr*

Energy to roast per gram roasted bean:

______ ÷ ______ = _______ *kW-hr/gram*

Total mass of beans used: _______ *grams*

Energy per gram × total grams used:

______ × ______ = _______ *kW-hr*

Roast B (bean type): ________________

Mass of green beans: ________ *grams*

Mass of roasted beans: ________ *grams*

Total energy of roast: ________ *kW-hr*

Energy to roast per gram roasted bean:

______ ÷ ______ = _______ *kW-hr/gram*

Total mass of beans used: _______ *grams*

Energy per gram × total grams used:

______ × ______ = _______ *kW-hr*

Total energy for roasted coffee actually used in your brew (Roast A + Roast B):

________+ ________ = ________ *kW-hr*

Brewing method(s): ____________________________________

Total mass of water heated: ________ *grams* *(make sure the kettle is cool and empty at start)*

Initial Water Temp: ________ °C Final Water Temp: ________ °C

Energy used to heat water: ________ *kW-hr*

Mass of brew: ________ – ________ = ________ *grams* *(filled carafe– empty carafe)*

RANDOM NUMBER GENERATOR: TARGET BREW MASS = _____ grams

Deficient mass: _______ – _______ = _______ *grams*

Energy penalty:_______ *grams* × 0.005 *kW-hr/gram* = ________ *kW-hr*

Total energy used to produce your coffee (roast energy + water energy + penalty):

________+________+ 0 = ________ *kW-hr*

Blind Taste Score: ____________ **Final Score:** ______________ *(kW-hr)* $^{-1}$

TDS Data for your Contest Brew

Total mass of all coffee grounds used for your contest brew: ________ *grams*

Mass of all brew in carafe: ________ *grams*

TDS: ________ %   PE: ________ × ________ ÷ ________ = ________ %

Part B – Measure TDS and submit your carafe and energy sheet

After you're done with the 1 L, take a quick sample to get the overall TDS and determine the PE. (Let your small sample cool to room temperature before you measure the TDS!) If you combined different brew types each might have a different PE, but this measurement will let you know what you ended up with.

As soon as you're done, bring both your filled carafe and completed energy sheet to the contest coordinator. After all the carafes are submitted, the carafes will be randomized by assigning a random letter known only to the contest coordinator.

Part C – The Blind Taste Test

After the randomized carafes are brought back out, it's time to taste. You'll have about 45 minutes to taste the coffee in each carafe. Everybody will taste every carafe, and each group will *collectively* decide on a score for each of the sensory attributes. (For example, your group will submit 12 scores for 12 carafes.) Refer to Lab 1 for a reminder on how each of the attributes is defined. If it's available, we recommend using the Cropster Cup app to record the taste scores… this allows rapid tabulation of the winners and the contest statistics.

For each tasting, dispense only about 1 ounce (half an espresso glass)… that's all you need. In fact, a few small sips of each is all that's necessary. Discard any remaining coffee. It's recommended that you drink some water in between coffees to help prevent 'taste fatigue.' Some people find it useful to bring some plain crackers to help cleanse the palate.

For our purposes, high numbers are always better. In the case of acidity, for example, don't give a high score if the coffee is unpleasant because it's extremely sour – that would be a low score. Likewise, a dull flat coffee without any acidity is also a low score. Give a high acidity score if the acidity is pleasantly 'bright' without being overwhelming. As soon as your group is done tasting, submit your score sheet to the contest coordinator, who will then go back and compile all the scores. You can clean up your station and take additional pictures or video while you're waiting. Before you leave, our volunteer will ring the gong again, and the winning groups will be announced. With good luck – and good engineering design – your group might take first place!

Design Competition Tasting Score Sheet

Group name: ____

Station number: ___________ **Section number:** __________

All categories are ranked 1 to 10 (1=terrible, 10=excellent), except balance, which is ranked –10 (gut wrenching terrible) to +5 (excellent).

Carafe	**A**
Fragrance	
Acidity	
Flavor	
Body	
Aftertaste	
Balance	
Total	

Carafe	**B**
Fragrance	
Acidity	
Flavor	
Body	
Aftertaste	
Balance	
Total	

Carafe	**C**
Fragrance	
Acidity	
Flavor	
Body	
Aftertaste	
Balance	
Total	

Carafe	**D**
Fragrance	
Acidity	
Flavor	
Body	
Aftertaste	
Balance	
Total	

Carafe	**E**
Fragrance	
Acidity	
Flavor	
Body	
Aftertaste	
Balance	
Total	

Carafe	**F**
Fragrance	
Acidity	
Flavor	
Body	
Aftertaste	
Balance	
Total	

Carafe	**G**
Fragrance	
Acidity	
Flavor	
Body	
Aftertaste	
Balance	
Total	

Carafe	**H**
Fragrance	
Acidity	
Flavor	
Body	
Aftertaste	
Balance	
Total	

Carafe	**I**
Fragrance	
Acidity	
Flavor	
Body	
Aftertaste	
Balance	
Total	

Carafe	**J**
Fragrance	
Acidity	
Flavor	
Body	
Aftertaste	
Balance	
Total	

Carafe	**K**
Fragrance	
Acidity	
Flavor	
Body	
Aftertaste	
Balance	
Total	

Carafe	**L**
Fragrance	
Acidity	
Flavor	
Body	
Aftertaste	
Balance	
Total	

Final Design Lab Report

For your final lab report, you will provide a "Design Report" describing your specific design logic and outcomes for the design contest. The following checklist is provided for your convenience.

Design Report Checklist

Report contents:

- ☐ Your group name, individual names, ID numbers, and section number (if applicable)
- ☐ Process flow diagram, with enough detail for others to replicate your process
- ☐ Mass balances, for water and coffee solids, including waste streams
- ☐ Type of beans and roast level(s)
- ☐ Energy usages for each step, and your logic for energy minimization
- ☐ TDS and PE for one of your final design brews (or competition brew)
- ☐ Your sensory evaluations of your brew and overview of scores from the blind tasting
- ☐ Summary of what you would do in a future hypothetical contest to improve

Finally, when you are done with your report, we recommend you sit back with a nice cup of freshly brewed coffee and reflect on all the engineering and science concepts you have learned about during this course. Hopefully you have a better sense now of how to think like a chemical engineer – and how to make excellent coffee!

Appendices

Appendix A – General Guidelines for Brewing

For those of you who are less experienced coffee brewers, the information below provides a quick and handy summary of "Coffee Basics" and "General Rules of Brewing." Coffee basics covers general information about getting the best results by starting with good quality roasted beans and water. The general rules of brewing are based on common taste perception and preferences. As you work through the labs and explore your own personal taste preferences, keep these general rules of brewing in mind in order to more rapidly optimize towards your own individual "nirvana in a cup." Happy brewing.

Coffee Basics

- Use freshly roasted whole beans: To make good coffee you need to use fresh beans. After roasting, the coffee is at its best a couple of days after roasting and up to two weeks if stored properly. Roasted coffee should be stored in an airtight one way valve bag. Note: do not brew beans for at least 1-2 days after roasting. Chemical reactions continue after roasting that are crucial for the flavor and aroma profile.
- Grind immediately before brewing: Grinding releases trapped volatile components. That great smell of freshly roasted coffee means you are losing those great flavor molecules rapidly, so use ground coffee immediately. Ground coffee becomes stale within a few hours of grinding. The smaller the grind, the more rapid the loss. For example, a fine espresso grind will be stale within a few minutes.
- Use good water: water is almost 99% of brewed coffee. The coffee stuff in your cup is only a little more than 1% of the total, so starting with good water is essential.
- Use the right "brew ratio" and the proper grind size: The right ratio of water to ground coffee and the grind size are based on the brewing method. The goal is to optimize the extraction of the good flavors without over-extracting the bad flavors.

General Rules of Brewing

- Ratio of ground coffee to water: For an automatic drip or pour over method of brewing a water to coffee ratio of 15 to 20 grams of water to every gram of ground coffee is typically recommended.
- Water temperature: The suggested water temperature range is 90-95°C (194-203°F).
- Grind size: A narrow particle size helps with an even brewing extraction and more consistent brewing results. The grind should also be optimized to the type of brewing method. For example, espresso uses a very fine grind while a French press uses a coarser grind.
- Extraction time: Typically, most brewing methods use 2-6 minutes (espresso is about 30 seconds). The goal is to extract 18-22% of the ground coffee mass into the brew to obtain approximately 1.15-1.35% total dissolved coffee solids in the brewed coffee. (You study this in more detail in Labs 6 and 11!)

Appendix B – Useful Units and Conversions for Coffee

Mass & volume conversions for water at room temperature

1 milliliter water = 1 gram (1 mL = 1 g)

1 liter = 1000 milliliters = 1000 grams = 1 kilogram (1 L = 1000 mL = 1000 g = 1 kg)

1 cup = 237 grams

1 cup = 8 fluid ounces

1 fluid ounce of water = 29.6 grams

4.25 cups of water = 1000 grams = 1 liter

Temperature

To convert from Celsius to Fahrenheit, multiply by 9/5 and add 32.
To convert from Fahrenheit to Celsius, subtract 32 and multiply the difference by 5/9.

°C	0	10	20	30	40	50	60	70	80	90	100	110	120
°F	32	50	68	86	104	122	140	158	176	194	212	230	248

°C	130	140	150	160	170	180	190	200	210	220	230	240	250
°F	266	284	302	320	338	356	374	392	410	428	446	464	482

Brew Ratio Table

Mass of Dry Grounds (grams)	Desired Brew Ratio: 14	15	16	17	18	19	20
5	70	75	80	85	90	95	100
10	140	150	160	170	180	190	200
15	210	225	240	255	270	285	300
20	280	300	320	340	360	380	400
25	350	375	400	425	450	475	500
30	420	450	480	510	540	570	600
35	490	525	560	595	630	665	700
40	560	600	640	680	720	760	800
45	630	675	720	765	810	855	900
50	700	750	800	850	900	950	1000

Mass of Water to Use (grams)

Notes

Notes

Made in United States
Orlando, FL
04 November 2026

90200501R00059